ELECTRONIC ELECTIONS

ELECTRONIC ELECTIONS

THE PERILS AND PROMISES
OF DIGITAL DEMOCRACY

R. MICHAEL ALVAREZ
AND
THAD E. HALL

PRINCETON UNIVERSITY PRESS

PRINCETON AND OXFORD

Published by Princeton University Press, 41 William Street, Princeton, New Jersey 08540

In the United Kingdom: Princeton University Press, 3 Market Place, Woodstock, Oxfordshire OX20 1SY

Library of Congress Cataloging-in-Publication Data

Alvarez, R. Michael, 1964–
  Electronic elections : the perils and promises of digital democracy / R. Michael Alvarez and Thad E. Hall.
      p. cm.
  Includes bibliographical references and index.
  ISBN 978-0-691-12517-6 (alk. paper)
    1. Electronic voting—United States.   2. Electronic voting—Security measures—United States.   3. Voting-machines—United States—Reliability.   I. Hall, Thad E. (Thad Edward), 1968– II. Title.

  JK1985.A484 2007
    324.6'5—dc22   2007008391

This book has been composed in Sabon

Printed on acid-free paper.∞

press.princeton.edu

Printed in the United States of America

10 9 8 7 6 5 4 3 2 1

CONTENTS

ILLUSTRATIONS

Figures

Tables

PREFACE

$\mathbf{W}$hen we started working on this, our second book project, we thought it would be relatively easy to get on paper our research on the electronic voting debate. After all, we have worked together for six years and we have written what probably amounts to thousands of pages on voting technology, election administration, and reform. Regardless of what we thought, this book proved to be a longer and more complex journey than we had planned—largely because the debates about the future of election administration in the United States have proved to be more political, more conflictual, and less predictable than we had anticipated.

We have managed to bring closure to this part of our electronic voting research. In the aftermath of the 2006 midterm elections, we are simultaneously thrilled by how much research we and others in this area have done in the past six years, but daunted by how much more work needs to be done. We are excited when policy makers and election officials listen to the research community and when they try to apply our research in their difficult work. When we look to the next few years, we see much more work that needs to be done but only a limited amount of time before the next major federal election cycle will be upon us.

We would like to thank the many people who have, directly or indirectly, given us their input, advice, or constructive criticism, as this particular project has gone from concept to final product. So many election officials, election reform advocates, voting system vendors, social scientists, and researchers have helped us that we will not attempt to name them all. We would, however, like to recognize those individuals we work with on a daily basis.

Various institutions and organizations have helped to support our research, financially and logistically. Carnegie Corporation of New York has supported the work of the Caltech/MIT Voting Technology Project since 2000 and gave us the financial support necessary to write this book and to undertake our unique survey research efforts reported in chapter 7 through a grant to the University of Utah. We also need to acknowledge the Dean's Office of the College of Social and Behavioral Science and the Department of Political Science at the University of Utah, which provided the initial funding for our survey research reported in chapter 7. We especially want to acknowledge the support and help of Geri Mannion, who provided us with a grant to support the writing of this book. Without her assistance, this book would not have been possible. Geri's support for election administration efforts has been truly a blessing for us and for the entire election community. Steve Ott and Ron Hrebenar at the University of Utah were also quite supportive of our work and agreed without hesitation to fund our initial study. The John S. and James L. Knight Foundation provided critical support of the Voting Technology Project work and that financial assistance helped us in many ways. Melissa Herrmann and her staff at International Communication Research were incredibly helpful in the conduct of our survey research, suggesting better ways of presenting the data and ensuring that the questions we asked about electronic voting addressed the issues we wanted to address. We are both involved in the Election Assistance Commission's Vote Count and Vote Recount Study, and this effort has been very educational for us, illustrating the reality of state election laws. We have also benefited from the interesting discussions we have had with the commissioners and the staff of the commission.

Rick Green at the University of Utah provided Hall with course release time on short notice during the initial writing of the book. While we were finishing this book, Alvarez was a Senior Fellow at the Annenberg Center for Communication at the University of Southern California; that fellowship provided interesting colleagues, great coffee, and much-needed quiet time for reading and reflection. Thanks to Jonathan Aronson, Geoffrey Cowan, and Simon Wilkie.

We both have had the unique honor of being associated with universities that give us great colleagues and students. At Caltech, we appreciate the input, good humor, and support we have received from our colleagues Jonathan Katz, Rod Kiewiet, and Peter Ordeshook. And without great students to help us in our research, and to challenge us as we talk about our research and perspective, our work would suffer: we thank Caltech graduate students Delia Bailey, Morgan Llewellyn, Betsy Sinclair, and Catherine Wilson for their input into our work. Delia Bailey and Betsy Sinclair assisted with some of the analyses we discuss in later chapters, and also commented on early versions of this manuscript. Caltech undergraduates Erin Hartman and Dan Knoepfle also provided helpful discussions and input into our work. At the University of Utah, we thank Erin Peterson, an undergraduate who has been an incredible resource helping us in our research. Thanks as well to our colleagues from the VTP, especially Charles Stewart. We also thank Monica Kohler, Cindy Brown, and Shelley Kruger at the University of Utah, who helped us manage the Carnegie grant and internal grants that supported our work. Finally at Caltech, we have to thank two wonderful people without whose daily support and assistance we would not have managed to get any work done over the past few years: the good cheer and incredible administrative support offered by

Karen Kerbs and Melissa Slemin have made our research efforts easier and more productive.

When we both started studying election administration in 2000 we had no idea where the journey would take us. Not only has our work been intellectually interesting—spanning a range of disciplines outside political science including law, sociology, public management, operations research, risk analysis, engineering, and media studies—but it has been a journey literally as well. We have attended numerous conferences on elections in Washington, D.C, and in other cities and have testified before Congress. We have observed U.S. elections and election administration separately or jointly in California, Utah, New Mexico, Oregon, Nevada, Texas, Virginia, Illinois, Washington, Georgia, and Colorado and Parliamentary elections in Estonia. Alvarez has observed elections in Argentina. Hall has given a talk on electronic elections in Estonia. Together, we have given talks in Switzerland, and at countless academic and election reform conferences or workshops. In all of these places, we have had experiences and met incredible people who have shaped our views on voting, election administration, and voting technologies. The quality of the academic work that is being done in the field of election administration has become much more sophisticated and nuanced over the past seven years, and we are the beneficiaries of this work. We are also fortunate to have been able to interact with international election administrators and social scientists, who have helped us to better appreciate the broader views associated with elections—beyond the narrow red state–blue state views often espoused in America.

The work we have done on our blog site on election administration—http://electionupdates.caltech.edu—has helped us keep up to date on the trends and events in elections. Paul Gronke, a great colleague who helps us with the blog, has been an asset in our work. We promise our readers that we will use the Election Updates blog site to keep them posted about developments and new research following the publication of this book.

Finally, we thank our spouses—Sarah Hamm-Alvarez and Nicole Hall—who have been exceptionally supportive of the time we have taken to learn about elections and the endless excitement we have about the vagaries of election administration. We dedicate this book to them.

Chapter 1

WHAT THIS BOOK IS ABOUT

Before the 2004 election, there was a blizzard of media coverage about the potential problems associated with electronic voting. Claims were made that the machines would lose your votes or would be hacked. Democrats and Republicans alike used these potential problems as a mechanism for mobilizing voters. For example, the Florida Republican Party sent out fliers in 2004 that said: "The liberal Democrats have already begun their attacks and the new electronic voting machines do not have a paper ballot to verify your vote in case of a recount. Make sure your vote counts. Order your absentee ballot today." Likewise, Democratic candidate Steve Henley was quoted on the campaign trail saying "By voting absentee, you make sure your vote gets counted. And in the event there is a close election, they have a physical copy of your vote."[1]

In 2004 voters in Broward County, Florida, were similarly encouraged to vote using absentee ballots so that they would not have to vote using the county's direct recording equipment (DRE) voting machines. By voting absentee, the voters were told that a paper record would exist of their vote and that it would be counted. Unfortunately, in the month preceding the November 2004 general election, as many as 58,000 absentee ballots in Broward County were lost after leaving the county election office.[2] Many voters there did not receive their ballot and could not easily vote any other way because their names were on the list of voters who had voted absentee. Moreover, it was expected that many of these voters would not receive the replacement absentee ballot in time for it to be returned and counted in the election. In an effort to use the debate over electronic voting to mobilize voters, thousands of voters may have been disenfranchised because the complexities of absentee voting had not been considered fully.

This story from the 2004 election illustrates a simple fact: life is full of risks, and all alternatives, including the choice not to act, carries with it inherent risks. This truism holds for elections as well, where all forms of voting carry inherent risks of problems, as a single procedural misstep can create an array of potential issues for voters. For example, the later the voters received the absentee ballots in Florida, the greater the likelihood that voters would return their ballot to the election office so late it would not be counted. Because these voters were now listed as absentee voters, they could not vote in a polling place or in early voting without bringing in their absentee ballot; otherwise, they would have to cast a provisional ballot. In any event, casting an absentee ballot carried its own potential problems. Even in the best of circumstances, in any election some percentage of absentee ballots are rejected because of voter errors either in completing the information on the absentee envelope or in missing the deadline for returning absentee ballots.[3] Also, absentee ballots maybe more likely to contain overvotes or undervotes compared to precinct-cast ballots, because absentee voters do not have access to the same convenient error-checking technologies that precinct voters can use today.

This book is about the risks and trade-offs associated with electronic voting. We consider how the media have framed the debate over electronic voting and how the public perceives this debate. Election reform is rarely considered through the lens of risk analysis and trade-off. Instead, reforms are attacked by various interest groups, who typically make claims about the risks of some method of voting. This is true not only in the area of voting technology, where debates rage between those who are concerned primarily with accessibility and those who are concerned about security. Reforms such as no-excuse absentee voting, early voting, vote centers, and even reforms to voter registration systems have all come under intense scrutiny, with claims made that such reforms will somehow negatively affect the electoral process or otherwise harm our democracy.

It has become common to consider election reform, and electronic voting, very critically and as a high-risk activity. In a provocative book about risk analysis, the legal scholar Cass Sunstein (2005) notes that much of the world is currently interested in a form of risk management known as the precautionary principle. This principle is based on the idea that the decision to mitigate a potential risk should not require the existence of absolute proof that it will come to fruition. Stronger versions of the principle have been expressed primarily in the context of environmental and health policy. According to the president of Friends of the Earth, for example, "the precautionary principle mandates that when there is a risk of significant health or environmental damage to others or to future generations, and when there is scientific uncertainty as to the nature of that damage or the likelihood of the risk, then decisions should be made so as to prevent such activities from being conducted unless and until scientific evidence shows that the damage *will not occur*" (Sunstein 2005, 193 emphasis added). As Sunstein notes, however, it would be difficult to meet such a high standard.

Sunstein also makes an important yet basic critique of the precautionary principle: it does not consider the risks posed by the status quo. To illustrate the point, he reports the advances that scientists have identified that would have been prohibited by the precautionary principle, including most vaccinations, open-heart surgery, x-rays, and antibiotics. One scientist identified "pasteurization, immunization; the use of chemicals and irradiation in crop variety development" as examples of items that would be banned by the precautionary principle. One could argue that AIDS research should be terminated because we do not know the risks, but because we know that the status quo is inherently dangerous (people die), we know that doing nothing will also cause, or continue, certain risks. Quite simply, while risks can produce failures, it can also produce great rewards.

In this book, we consider the risk of electronic voting in light of what we know about the status quo. We begin by examining various frames that have been used to express the risks associated with the election process.

THREE FRAMES FOR CONSIDERING ELECTIONS

For thirty-seven days in November and December 2000, while America waited to learn who would be the next president of the United States, election officials were the butt of late-night television monologues and were vilified in the media. The spotlight shown brightly on this one aspect of our political system, and people around the world learned about what had previously been an esoteric subject—American election administration. During this period, people also developed very distinct impressions about the conduct of elections, and three divergent views of the election took hold. These views reflect the nature of how people frame events for political and social purposes and how such framings are then used to discuss the risk of similar problems occuring in the future.

The first frame was that the election was a failure of administration across the electoral process. Various problems occurred in the election—from registration problems to voting system failures to improper poll worker actions—but they occurred because elections historically have been neglected. Fortunately, failures of administration are something we know how to deal with in America: we create commissions! And create them we did. There was a National Commission on Federal Election Reform that was chaired by former presidents Jimmy Carter and Gerald Ford and supported by scholars from several elite universities. An academic commission—the Caltech/MIT Voting Technology Project—provided an intellectual basis for election reform. Many states set up committees or commissioned reports to study election processes in their own states; for example, in Florida Governor Jeb Bush appointed a twenty-one-member Governor's Select Task Force on

Elections Procedures, Standards and Technology immediately in the wake of the *Bush v. Gore* ruling by the U.S. Supreme Court.[4] Every relevant interest group—from the secretaries of states to the House Democratic caucus—also created a commission or task force and issued a report. The culmination of this work was the passage of the Help America Vote Act (HAVA) in 2002, which injected several billion dollars into the electoral process and is now reshaping the way in which elections are conducted.

A second frame was that the election illustrated that voting is a civil right, and on this score the election failed large segments of the population. Especially since the passage of the Voting Rights Act in 1965, which sought to strengthen Section 1 of the Constitution's Fifteenth Amendment ("The right of citizens of the United States to vote shall not be denied or abridged by the United States or by any State on account of race, color, or previous condition of servitude"), the ability of citizens to vote has been viewed as an important civil right. And while administrative failures lead to commissions and lawmaking, civil rights failures often lead to lawsuits and administrative remedies. Not surprisingly, then, numerous lawsuits were filed after the 2000 election challenging various aspects of election administration, but especially the voting equipment used in 2000 and the civil rights failures some of this equipment caused.

In California, the American Civil Liberties Union sued the state over the use of punch cards, while in Georgia it sued over the use of punch cards and optical scan equipment, as both types of equipment were argued to produce racial disparities. In several cities, organizations representing people with disabilities sued over any systems that did not provide an interface that would allow people with disabilities—especially the blind and people with very limited motor skills—to cast ballots without assistance. This view of voting as a civil right also is important to social scientists and many in politics, but in a slightly broader manner. Specifically, voting is a key part of citizenship and, as such, should be strongly encouraged. As both President Carter and President Clinton said after the 2000 election, it should be easy to register, easy to vote, and easy to count the votes. Making it easy for people who have historically been marginalized in the voting process benefits everyone. As we write this book, litigation over voting machines and election practices has become common in the United States.

The third frame is that the 2000 election was a fraud; the election was stolen. People on both sides of the political spectrum still hold this view. People on the left believe that Republicans stole the election in Florida, while people on the right point to places like St. Louis, Missouri, as an example of fraud in the works. This view that fraud is rampant—or potentially rampant—is not new. The very existence of voter registration laws, including requirements that voters register as much as thirty days before an election, are but one example of a policy designed to deter fraud. In general, those who view the electoral process with a strong concern about fraud are likely to have a view contrary to that held by presidents Carter and Clinton: elections should be designed to thwart fraud, even if it makes it difficult for some people to vote.

These three frames can determine the way in which policy makers and the public view the risk that a future election crisis will occur. After the 2000 election, the election reform debate centered primarily on the first two frames. Election reform required improving election administration and ensuring that all citizens were provided with the opportunity to participate in a meaningful way in the electoral process. In fact, most election reform commissions in 2001 specifically avoided the topic of fraud and who was responsible for the problems in the 2000 election. Instead, the focus was on improving the electoral system and making the system work well for every person in America. HAVA addressed the concerns raised by the reform commissions and created a process for moving election reform forward in the states. HAVA also opened the door for the federal government to provide substantial funding to states for the purchase of new voting equipment.

Since the passage of HAVA and in the wake of the recent 2004 election, however, this view of elections as a civil right to be well administered has been overtaken by a view that the potential for fraud or glitches has become rampant. This view is especially held by some computer scientists and others who oppose electronic voting. Specifically, they argue that a potential for fraud exists because of the use of computers in elections for voting. According to these critics, direct recording electronic voting equipment, which has become popular because of its ability to enfranchise historically disenfranchised voters, are "black boxes" that are likely to contain malicious code that steal votes and steal elections. In essence, because hacking a DRE may be theoretically possible, it is inevitable that such hacking will occur. These critics are pushing efforts in Congress and states across the country to stop the deployment of DREs—with lawsuits and legislative initiatives—and to force states either to move back to paper systems or to outfit DREs with printers that will print a paper "receipt." Some even oppose efforts to create electronic audit trails, using state-of-the art cryptographics.

This book concerns, the ongoing debate about how Americans will cast ballots in future elections. We have had this debate throughout the history of the United States. Consider the evolution of voting rights since the nation's founding. Most people assume that the history of the United States is one where the franchise has been systematically enlarged. First nonlandowning males were granted the right to vote. Then the Fifteenth Amendment gave African Americans de jure voting rights—with de facto rights coming only with the passage of the Voting Rights Act—and the Nineteenth Amendment gave women voting rights. Finally, the Twenty-sixth Amendment gave individuals between the ages of eighteen and twenty-one the right to vote, making America a more fully democratic society. As many scholars have noted, however, this history is incomplete. During our nation's history, we have had periods of legal expansions of voting rights, but also periods where the law has been used—intentionally or unintentionally—to disenfranchise.

Many of these periods of marked disenfranchisement have occurred because of concerns about election fraud. For example, Keyssar (2000) notes that literacy tests, poll taxes, and voter registration were all designed to limit election fraud. Although we typically associate some of these tactics with efforts to disenfranchise African Americans, they did have "progressive" proponents who did not think it appropriate for those with limited educations or who owed debts to the state to be able to participate in elections. For example, the *New York Times* (1923) referred to that state's literacy test as "a wholesome law" and supported the legislation creating the law. Progressive "good government" groups argued that voters who could not pass the literacy test could be easily swayed, have their votes purchased, or make uneducated choices. Even efforts that seem reasonable today, such as voter registration, were originally quite onerous because of the way it was implemented. A voter typically had to register in person during office hours (between nine and five o' clock) at some central government office, and do so annually. This effort kept the voting rolls fresh and limited the ability of fraudulent voters from being on the rolls. The fact that it made it difficult for many otherwise eligible citizens to stay registered, thus effectively disenfranchising them, was the price of stopping fraud.

Today, the conflict over the efficacy of electronic voting machines has put two of the visions of elections—as public administration and as a civil right—in direct conflict with the third vision, a strong concern about fraud or glitches. Unfortunately, this conflict has real implications for real people. Because newer versions of electronic voting machines can help voters minimize simple errors, like overvoting or the unintentional skipping of races on the ballot, these newer voting machines might be significantly more accurate than their exclusively paper-based predecessors. Electronic voting machines can also take some of the subjectivity out of vote tabulation; again, by making it easier for voters to avoid simple mistakes, electronic voting machines could make it easier to assume that the ballot choices are consistent with the voter's intentions.

Without electronic voting, many people with disabilities and voters with limited English proficiency are effectively disenfranchised because they literally cannot cast a ballot using a paper-based technology. They can ask someone to vote for them, but they can never be sure their vote was cast correctly and consistent with their intentions. Likewise, data from Georgia and other states show that paper-based voting also can result in disparate results between white and African American voters. A recent study of Georgia's transition to statewide electronic voting found that voters in predominantly white communities were more likely to have their votes counted when using paper-based voting technologies compared to voters in minority communities but that these disparities were reduced when the state shifted to electronic voting machines.[5]

But, electronic voting machines have potential drawbacks. Because these computer-based systems are modern and complex devices, their electronic designs make it more difficult to spot simple glitches in the hundreds of thousands of lines of computer code that run these machines, let alone deliberate attempts to insert nefarious computer code into their innards. In the current marketplace, electronic voting systems are expensive, sometimes costing thousands of dollars per voting unit. Even if cost is not an issue, the transition from existing "legacy" systems to the new electronic voting systems requires that election officials juggle a complicated integration of modern computerized voting machines with more traditional election administration systems, unless they replace these other election administration systems completely. Finally, a lack of transparency associated with many aspects of the testing, certification, and use of these electronic voting devices continues to fuel questions and concerns about how electronic voting machines "really" work.

In this book, we examine whether the United States should transition to electronic voting and what forms of electronic voting should be allowed. Readers should understand that both sides in this debate have strong arguments and claims that merit attention, a point we make repeatedly in the early chapters of this book. But our thesis is that these claims can and should be subjected to scientific analysis and that this debate about how our democracy should conduct elections must be settled by testable hypotheses, real facts, and empirical analysis—not political rhetoric. The latter part of this book provides a framework that can be used for just this type of scientific study of electronic voting systems and presents a variety of studies of different applications of electronic voting, both in the United States and abroad. These studies provide helpful data that will assist in public deliberations about the ways in which electronic voting can be adopted in the United States.

WHAT DO WE MEAN BY ELECTRONIC VOTING?

Before delving any further into our argument, we need to be clear on terminology. We have found in recent years that there is substantial confusion over the exact nature of the technology of elections in the United States, as many otherwise well-informed people (including many supposed experts on election administration) are sometimes surprised to learn the extent to which electronic technologies have come to dominate the overall process of American election administration.

First, although most of the public scrutiny of the election process occurs in the days before and after major national elections, substantial work must be done in the months—even years—before and after an election, even in what might appear to be relatively simple municipal contests. In major national elections, the preparation and logistic enterprises required are very much like a major military mobilization. The process of election administration includes the geographical division of the jurisdiction into registration and voting precincts; the acquisition, maintenance, and storage of devices used by voters to cast their ballots; the development and testing of ballot definitions and designs; the development and distribution of voter educational materials and outreach campaigns; the designation of locations for early and election day voting stations; the recruitment and training of workers for early, election day, and postelection canvass activities; voter registration updating and purging; planning of absentee voter applications, authentication of voters, distribution and receipt of absentee materials; early and election day voting activities; election night auditing and receipt of ballots and voting materials; initial tabulation of results; resolution of disputed and problematic ballots; postelection canvassing and final reporting; mandatory or discretionary recounts; and other postelection administrative actions. Today, many—if not most—of these election administration tasks, in many places, are conducted using electronic devices and computerized technologies.

But our analysis in this book concentrates mainly on the technologies used by voters to cast their ballots—typically in precincts on election day, but increasingly before election day through early or absentee voting. Thus, when we say "electronic voting," we really do mean to concentrate attention on the act of "voting"; although new technologies are increasingly being used for many of the activities of election administration, those are not our concern here. This distinction is critical and often lost in the debate over electronic voting. For example, a recent article in the *Tampa Tribune* discusses "the need for a strong examination of Florida's electronic voting system"; however, the system in question is an electronic voter registration system, not a direct recording electronic voting machine.[6]

Second, what do we mean by "electronic"? A century ago, many citizens cast their vote using paper ballots, which were then counted by hand. Others used what were then called "voting machines," mechanical devices that we now tend to call "lever machines." Votes cast on those mechanical devices were recorded through some type of mechanical device, sometime like an odometer in a car, and vote tallies were then read from the mechanical tally device by election workers at the close of voting. We briefly cover the history of voting systems in chapter 2.

While some Americans still cast votes using hand-counted paper ballots or mechanical voting machines, the number of ballots cast in recent elections using these old technologies has diminished dramatically. In their place are two conceptually distinct voting technologies. One of these voting technologies involves marking a paper ballot, which is then tabulated by an electronic device. This technology has two categories: the punch card ballot in which voter preference is indicated by making a hole in the ballot; and the increasingly common optical scan ballot in which voter preference is indicated by filling in the circle next to a candidate's name or completing an arrow pointing to a candidate's name. Both punch card and optical scan ballots are then tabulated by an electronic device, though they can also be (and sometimes are) counted by hand.[7]

In our definition, an electronic voting device is one in which the voter inputs preferences electronically—either flipping some mechanical levers that record a vote into the electronic voting device (the so-called direct recording electronic device), tapping selections on a "touch screen" voting system, or using some other input method to indicate a vote to an electronic voting device. When using electronic voting technologies, the voter is interacting with a computerized system that translates the voter's input into an electronic stream of information that is then somehow recorded and preserved for later tabulation. The electronic voting machine might simply record the voter's input into some type of electronic storage device or devices (involving nonremovable or removable media), it might translate the voter's input onto a paper ballot that is printed for the voter to verify and deposit in a ballot box, or it might store the voter's input electronically and provide a printed ballot that the voter can verify. As long as the voter's preferences are being recorded, by the voter, into some initial stream of electronic information, we consider that to be electronic voting. Later, in some places where the details are relevant to our discussion, we sometimes differentiate between different types of electronic voting machines.

The electronic voting device can stand by itself, be networked to other voting devices in a precinct or early voting station (local-area network

[LAN]), or be attached to a wide-area network ([WAN], like the Internet). In the present discussion, we maintain the distinction between WAN electronic voting, which we will call "Internet voting" (and which we have discussed extensively in our 2004 book on the topic), and stand-alone and LAN electronic voting.[8] The distinction follows the typical parameters of public debate about WAN versus non-WAN electronic voting.

OVERVIEW OF OUR ARGUMENT

In analyzing the political and policy implications of the current debate over electronic voting, we focus especially on the issue of risk assessment, the regulatory framework under which voting technologies operate, and the need to hold all voting systems—both electronic and paper—to high standards. The book can be divided into three parts. We begin by examining the arguments and data supporting both sides of the debate so that readers can understand their competing arguments and claims. In chapter 2 we trace the evolution of voting technology in the United States in order to provide readers with a brief historical context in which to understand the current debate over electronic voting. The second half of this chapter focuses on how electronic voting could revolutionize participation among voting populations that have historically been disenfranchised. In chapter 3 we examine the claims of critics of electronic voting. We start by putting the concerns of these critics into the theoretical context of the risk society. We then examine the explicit claims of these critics regarding the security risks associated with electronic voting.

Next, we examine the politics of the debate over electronic voting, presenting data on the political framing of the debate, and how this framing compares with the data from elections across the country and the perceptions of the general public. In chapter 4 we examine the role of interest groups and the media in the framing of the electronic voting debate. After the 2000 election, the general view of electronic voting was positive, with the media expressing concerns about the paper-based systems that failed in Florida. Starting in mid-2003, the media story shifted, with electronic voting viewed as being a key potential source of fraud and bias in the electoral process. In chapter 5 we review how the debate on electronic voting spilled over into the ongoing efforts to experiment with Internet voting. This spillover resulted in the termination of the Department of Defense's SERVE project (Secure Electronic Registration and Voting Experiment), although there was a successful Internet voting trial conducted in the Michigan Democratic caucus. In chapter 6 we review the research on the successes and failures of electronic and paper-based voting from 2000 to 2006. This chapter provides a foundation for understanding what science shows to be factual—not merely rhetorical—in this debate. In chapter 7 we present unique data on the public's acceptance of electronic voting. Two national surveys are presented—one conducted before the 2004 election, one conducted after—to determine how voters view this new technology.

Finally, we present a risk assessment framework that can be used for the scientific study of electronic voting systems, a regulatory framework that can be used to move electronic voting forward, and an implementation framework for successfully moving to electronic voting in jurisdictions across the nation. In chapter 8 we develop a process for implementing these various risk assessment models, which includes identification and analysis of threats through a threat-risk assessment, mitigation of threats through procedures and design, implementation of a system within the context of the model, collection of forensic data on implementation, and updating of threat-risk assessment on the basis of the forensics and changes in the implementation environment. Finally, in the conclusion we provide ten recommendations that we think can move the debate over electronic voting forward.

Throughout this book, we stress that the claims on both sides of the debate can and should be subjected to scientific analysis, and that this debate must be settled by resort to facts and evidence, not political rhetoric. Our goal is to let the data, not any preconceived set of views or biases, speak. In chapter 4 we discuss how electoral politics has become absurdly caricatured through the simplistic view that there are "red" and "blue" states or "red" and "blue" Americans. Likewise, the debate over electronic voting has been caricatured as a battle between parties, ideologies, ethics, and values. This debate needs to become more rational, and this book is our attempt to move the debate in that direction.

Chapter 2

PAPER PROBLEMS, ELECTRONIC PROMISES

We commonly have people ask us, What is so difficult about running an election? Having heard this question so many times, we have a good guess that the next question is: Why can't we just vote on paper ballots? Sometimes, people do surprise us, like twenty-seven-year-old Catherine Getches from Santa Rosa, California, who wrote the *Los Angeles Times* to ask why we vote in such an antiquated way. Like many others, she expresses frustration with the way we are forced to vote:

> It was Super Tuesday when I realized that the Urban Outfitters' tee that had at first annoyed me was actually true. "Only Old People Vote," the shirt declared in my generation's ironically detached way. It was true—voting made me feel more of a nonmember than a participant in the political process. I cast my ballot in one of the area's oldest elementary schools, where I signed in with an election volunteer wielding a magnifying glass. And when she had trouble finding my name she joked about alphabet amnesia, having "learned the ABCs so darn long ago." I took a ballot from a man sipping from a can of Ensure, and I was handed a dried-up Sharpie by a silver-haired woman who kept the cap as insurance on its return.
>
> I wondered if the word "primary" was a nod to the primitive voting system and if these three attendants were at this pioneering polling place when it all began. I pictured my paper ballot racing to the counting location via Pony Express in time for that night's results. And for the first time, I wished life was more as it was on "American Idol." If vast numbers of people can be motivated to vote for contestants who are simply interested in being heard, why is it so hard to get people to act in behalf of candidates who have a message?
>
> If I can transfer money via phone, publish photos straight from my cell phone to the Web and instant-message a vote along with millions of others to elect an Idol, why can't there be a safe and modern system in place for selecting the country's leaders? And why do we think the current system is so safe? At my polling place, no ID was required, just knowledge of my name. My privacy was subject to the voice level of the attendants, and the cereal box-like contraption that was to conceal votes from wandering eyes was hardly fool- or snoop-proof.[1]

The questions raised by Getches are typical of the debate about how Americans cast their ballots, and they suggest a dissatisfaction with both old and new voting technologies. This is due in part to a poorly informed perception of the process of holding an election and a lack of understanding of the complexity of the American electoral process.

Even among critics of how Americans vote, few have really stopped to consider the complexities of the election process and even fewer have taken the opportunity to actually study the breadth and depth of election administration issues in the United States. Underlying these questions about election are views of risk, security, and preferences for specific technologies that may or may not be based on the full range of facts and knowledge we currently have.

We can unravel this complexity by examining briefly the history of voting technologies in the United States. This historical digression can be divided into three parts: the first 120 years of the Republic, beginning with the movement from oral voting in public to the use of party-distributed paper ballots to the secret paper ballot; roughly the first half of the twentieth century, which saw the introduction of new voting systems, such as lever machines; and the recent period, with the introduction of punch cards, optical scan ballots, and electronic machines. Interestingly, throughout this history there have been claims that new voting technologies would result in increased levels of fraud and calls that elections would be "stolen" as a result.[2]

THE "GOOD OLD DAYS"

American elections have a colorful and vibrant history, a lost past that many observers of contemporary elections lament. Gone are the torch-lit parades of the nineteenth century, days when the streets would literally be full of loud partisan debate. Likewise, political parties today choose their nominees (largely) in public primaries, not through brawls where the winning party faction literally beat—and threw out the door—members of the losing faction. Instead of a vibrant public spectacle, we have sterile and staged televised debates, where candidates compete to provide the best sound bite for the late evening news and rarely if ever look at anything but the camera lens or the moderator's face. Gone are the great orators of the past, candidates or politicians who could speak to the masses and simultaneously elaborate an issue platform and rouse citizens to the cause. Instead, we have candidates who communicate using electronic formats—television commercials and increasingly the Internet—with brief bursts of negativity that may serve to turn voters off to politics.

Gone also is the simple act of voting, the literal counting of hands for one candidate or another. Instead, we have voters who are asked to decide dozens of candidate races, and sometimes many dozens of complex issues, all on a single ballot. Gone are the good old days, the simple days of early American political history (at least as they are remembered by many critics of the contemporary political scene).[3] Even in those days when we did ask voters to cast a more complex ballot containing multiple candidates, the activity was actually quite easy. Individuals who worked for political parties or political factions distributed complete ballot slates to voters, which the voter then cast in *toto*. Today, voters are frequently called upon to vote for dozens of candidates (many of whom they may know little about) and then consider numerous poorly worded constitutional amendments or ballot initiatives, which can make voting a cumbersome and time consuming task.

Drawing broad generalizations about the history of the American political process is difficult, and much of that difficulty arises from the simple fact that election practices and procedures have always varied considerably across the states. Despite these differences, we see three broad historical periods that characterize the basic parameters by which Americans have voted since the founding of the nation: the partisan paper ballot days (roughly running from the nation's founding to the late nineteenth century); the nonpartisan paper or machine voting days (from the late nineteenth century through the early 1960s); and the advent of electronic technologies in the voting process (running from the early 1960s to the present).

Although we use the term "paper ballot" in the partisan paper ballot days, in some places early in our nation's history voice voting was actually used for public and political elections (well into the post–Civil War era).[4] In this time, the election process was relatively simple and largely controlled by political parties. Usually a voter would obtain a ballot that listed all of the party's candidates then running for office from a party official, friend, or family member or sometimes from a newspaper. At this time, the voting technology—the prevoted paper ballot—was often supplied not by impartial election officials but by partisans.

Armed with this paper ballot, the voter would take it into a polling place on election day and drop it into the ballot collection box after authenticating his voting eligibility. In many cases, these paper ballots were large and colorful, and even ornamented with party labels and slogans.[5] The idea was that the bigger and more colorful, the easier it was for party officials to monitor who was casting ballots for their party's candidates. As voters were carrying their ballots openly into the polling place (sometimes literally obtaining the party ballot from the party official on the street in front of the poll site), the ballot during this period was not secret. Voters had to go out of their way, sometimes at great physical cost, either to obscure which party they were supporting or to alter the partisan ballot in some way so as to not cast a straight party ballot.[6]

The lack of a secret ballot during this period enabled party officials to engage in various schemes to corrupt the electoral process. Party officials could easily (and sometimes quite cheaply, just for the price of a few drinks) purchase votes and efficiently monitor whether those votes were cast as purchased. For these voters, the loss of secrecy was compensated for by the gratuity the voter received from the political party, something that many voters were more than willing to accept or even negotiate. Voters could also be coerced—either physically harmed if they were not carrying the "right" party ballot into a polling place, or threatened with physical harm if they did not vote "correctly." Party officials could also engage in less overt attempts to buy votes, or to coerce voters, through employers. All of these schemes were possible during this early period of American history due to the public nature of balloting.[7]

Voting has not always been associated with a ballot. In the late 1700s and early 1800s in the South, voting had an oral tradition. Men would gather before an election judge on election day and, when their name was called, state verbally which candidate they supported. An alternate version of this was used in Virginia, where men would cast a vote by writing their name in a poll book next to the name of their preferred candidate. In both cases, this election system explicitly linked a voter to a candidate and could ensure that there was no voting fraud via ballot box stuffing.[8] The process was completely transparent, but at the cost of potential voter intimidation or coercion.

The problems of voter intimidation and coercion associated with the lack of secret ballots led to two important innovations in election administration, reforms that swept the nation in the late nineteenth and early twentieth centuries. This began the period of the nonpartisan paper or machine voting days, a time of significant political and electoral reform in most states. States changed their electoral procedures to adopt a secret, multicandidate ballot (the so-called Australian ballot) to discourage vote buying and electoral corruption. Also, many of the larger urban areas (especially along the East Coast), where partisan political machines had long held control, moved away from the use of paper ballots to mechanical vote casting and recording devices. These new voting devices were simply referred to as "voting machines." Today, some of these voting systems are still in use, though they are more commonly known now as "lever voting machines."

The machine voting devices, or lever machines, have a common voter interface. The voter usually stands in front of a large set of small levers (sometimes pulling a curtain closed behind him). Next to each small lever is a candidate's name, or the description of a ballot measure. Sets of levers are grouped according to which race they belong with.[9] In many locations, these machine voting devices allowed for the casting of a straight party ballot; the voter simply threw one (often much larger) lever to cast a complete partisan ballot for candidates of one party or the other. Each vote choice was then recorded mechanically, often with the use of physical counters, much like the mileage counter most of us are familiar with in our cars.

These mechanical voting devices were intended to alleviate problems associated with the use of paper ballots, the most significant problems being fraud and corruption. The large lever machines were considered to be harder to steal, harder to stuff, and harder to throw into the river than paper ballots. However, the mechanical devices themselves were no panacea to fraud and corruption, and they introduced a series of additional problems that may not have been easily predicted when they were adopted.

Mechanical voting machines are not impervious to voter coercion or election fraud. Voters can be forced to show an observer the levers they have thrown before they cast their ballot, either by the simple means of leaving the ballot curtain open or by allowing an observer into the voting booth with them.[10] Other strategies can be used to manipulate a mechanical vote—for example, using a foreign object to make certain levers difficult or impossible to move, which that can effectively disenfranchise the potential supporters of some candidates relative to others. Also, individuals with access to the mechanical vote storage device can manipulate the stored vote totals—before, during, and after the election. Once the vote totals have been read from the storage unit and the counters reset to zero, there is no way that the ballots can be effectively recounted, meaning that vote totals can be misread (by error or deliberately) from the mechanical device.

Mechanical voting machines can be very difficult for some voters to use. If the entire ballot is to be presented to a voter (a "full-face" voting device), the voting machine might be the size of a large refrigerator. The sheer size of many of these mechanical voting machines means those voters with physical handicaps, who are vision impaired, or who are shorter than the voting machine may have significant physical problems using mechanical voting machines.[11] These voting devices cannot easily render ballots in multiple languages. Finally, mechanical voting devices can have ballot designs that render it difficult for voters to find certain races or lead them to inadvertently miss races that are on the ballot.[12]

Still, while some states and counties turned to mechanical voting devices during this period, most of the rest of the United States continued to rely on paper ballots for their elections. Increasingly, paper ballots produced problems for election administrators and voters. After the Progressive reforms at the beginning of the twentieth century, in many locations ballots lengthened considerably as voters were called upon to elect representatives to scores of state, regional, and local positions. Voters were also called upon to decide matters of policy in the voting booth as well, with the advent of the referendum, recall, and initiative process in many states (especially the western states).

With the increasing length of the ballot came a variety of legal and procedural constraints, especially in the latter half of the twentieth century. This led election officials to seek alternatives to paper ballots and mechanical voting devices. For example, as awareness grew that there were potential "ballot order effects" (where candidates listed first or last on the ballot are argued to receive marginally greater numbers of votes than had they been listed in other ballot positions, especially in less salient contests), election officials needed ways to produce multiple "types" of ballots, where candidate names in a specific race in a jurisdiction would be randomly ordered (and sometimes rotated to different places on the ballot) to alleviate these potential problems (Merriam and Overacker 1928; J. Harris 1934; Bain and Hecock 1957). Also, after the passage of the 1975 amendments to the Voting Rights Act, election officials in many parts of the nation had to provide ballots in multiple languages. In many places population growth simply made the use of mechanical voting devices and paper ballots impractical—mechanical devices had to be purchased, maintained, and transported to hundreds (and in some cases thousands) of precincts; hundreds of thousands (and sometimes millions) of paper ballots needed to be printed, distributed, accounted for, and then counted by hand. Election officials needed ways to deal with this complexity more efficiently and to handle the sheer magnitude of the balloting problem, and during the latter half of the twentieth century they turned to electronic technologies to resolve many of these problems.[13]

Thus began the advent of electronic technologies in the election process. In the early 1960s the first wave of new voting systems, using punch cards, swept across the nation. Many large election jurisdictions moved to adopt punch card voting systems, primarily to provide a cheaper and easy-to-administer voting system in large, urban areas. Two different varieties of punch card voting systems were developed and used in the United States. The first variety, which usually goes by the trade name Votomatic, uses a thin computer card with columns and rows of very small, prepunched rectangles. The voter takes this card and inserts it into a voter recorder device; the device has a booklike representation of the ballot, and when set up correctly, highlights the appropriate hole that the voter should check to cast a vote. When done, the voter pulls the card out of the device, and then inserts it into a ballot box. Typically, the punch cards are taken to a central location where they are read and tabulated quickly, using large punch card readers.

The second primary type of punch card voting system is commonly called Datavote, and it differs from Votomatic in two regards. First, the names of candidates or ballot measures are printed directly upon the card, and the card is inserted by the voter into a device that looks like a large hole-punch machine. The voter moves the hole-punch handle until the device highlights the choice the voter wishes to make, at which time the voter pushes on the handle; the device then punches a rectangular hole in the card. The second primary difference between Votomatic and Datavote is that the Datavote system has no prescored punch-out holes. Instead, the Datavote recorder device creates the hole-punch for the voter.[14]

Punch card voting systems, as they have typically been used, are praised by election officials because they are argued to be cheaper and more cost-effective than other voting systems, especially in very large election jurisdictions with complicated and complex ballots (for example, Los Angeles County has until recently been a longtime user of the Votomatic system). But punch card voting systems have many detractors, with problems noted by observers of voting technology well before the 2000 presidential election.

For example, Roy Saltman, in his 1988 study of computerized vote-tallying systems, noted a series of "vulnerabilities" of the Votomatic and Datavote punch card systems. Regarding Votomatic, Saltman pointed to lack of candidate information on the punch card, the need to turn all of the pages in the vote recorder, alignment failures, and malicious alteration of the instructions. With Datavote, Saltman noted other "vulnerabilities," including the need for multiple cards and the difficulties of processing multiple cards. Clearly, problems of punch card voting systems had been well documented and publicized in the election community long before the 2000 Florida punch card voting debacle.

Punch card voting systems experienced other problems, beyond those noted by Saltman and other earlier studies of voting technology. One of these important problems is the inaccessibility of these voting devices for voters with disabilities: voters with vision impairments find it difficult to see the ballot and punch card, and voters with physical impairments find it physically difficult to use the stylus to punch out the Votomatic chad or the handle to make their mark using Datavote. The best option that has been developed is for these voters to have assistance, from either a polling place worker or some other individual (usually a friend or family member).

But "assisted voting" raises an array of concerns, especially coercion and the general inability of voters using assistance to cast a secret ballot. Two examples illustrate this point. Blind individuals can vote using punch cards only by having someone read them the choices and then voting the selected candidates. However, this forces blind individuals to trust that the ballot is being marked as requested and puts these voters in the position of having their vote choice overheard by others. Likewise, elderly individuals with limited manual dexterity can similarly be forced to have someone in the voting booth with them who would know how they voted.

Critics of punch card voting systems point to a variety of studies that document their relative error rates, especially the differences in residual vote rates, between punch card and other voting systems. The residual vote, which is typically defined as the difference between the number of votes counted in a specific race relative to the number of ballots cast in the election, is a commonly used metric of voting system accuracy (Caltech/MIT Voting Technology Project 2001; Ansolabehere and Stewart 2005; Alvarez, Ansolabehere, and Stewart 2005; Hamner and Traugott 2004). Studies in the wake of the 2000 presidential election commonly noted the relatively high residual vote rate in jurisdictions using punch card voting systems. For example, the Caltech/MIT Voting Technology Project (2001) found that the presidential residual vote rate for punch card voting systems was 2.5 percent, compared to 2.3 percent for electronic voting systems, 1.8 percent for paper ballots, and 1.5 percent for both optical scanning and lever machines. The high residual vote rates of punch card voting systems, combined with concerns about the inaccessibility of these voting devices for many voters with disabilities, led to calls for their abandonment.

Other problems with punch card voting systems have been studied recently. One intriguing problem was found in the wake of the 2003 California recall election, the last major statewide election in California involving the use of prescored punch card voting devices (as the result of litigation, prescored punch card voting systems were decertified for use in California in 2004). Research reported by Sled (2003) and Alvarez et al. (2004) found a "vertical proximity" effect (also called an "adjacency effect"), where candidates who were lucky enough to be placed on the ballot in close proximity to prominent candidates (for example, George Schwartzman, who was next to Arnold Schwarzenegger) received a statistically higher number of votes than would have been expected had they not been listed next to a prominent candidate on the recall ballot. Furthermore, these "vertical proximity" effects were statistically greater in California counties using punch card voting systems.

Additional studies have documented other problems with punch card voting systems, especially the Votomatic punch card voting system.

A series of academic studies published in the wake of the 2000 election controversy, noted differences in residual votes, overvotes, and undervotes that raised concerns about voting rights violations (Alvarez, Sinclair and Wilson 2002; Buchler, Jarvis, and McNulty 2004; Sinclair and Alvarez 2004; Tomz and van Houweling 2003; see also Ansolabehere 2002). These studies tend to document that punch card voting systems were often used in election jurisdictions that had high minority populations, and also that minority voters appeared to be more likely to overvote and undervote when using punch card voting systems (especially prescored punch card voting systems) relative to nonwhites. Studies like these led to efforts at the state level (for example, Georgia and California) and the federal level (the federal Help America Vote Act) either to ban punch card voting systems or to lead counties towards the use of other voting technologies.

The problems with punch card voting systems, especially those seen in the 2000 presidential election in Florida and the studies that came in the wake of that election, have led to a marked shift in voting equipment use by the 2004 election. Some states, like Georgia, have moved entirely from a mélange of voting technologies (virtually every type of voting technology available was used in Georgia in the 2000 presidential election) to the exclusive use of one electronic voting system. Other states, like Florida, has shifted from a similar mélange in the 2000 election to either electronic voting systems or optically scanned ballots. And the third set of states (California, for example) has moved away from punch card voting systems but has not had a consistent state-led effort to adopt particular voting technologies.

Table 2.1 shows the evolution in the use of voting technologies in the United States since 1988. It shows a drop in the use of paper ballots and lever machines by American counties since 1988; these voting technologies were used in the vast majority of American counties in 1988 but were rarely used by 2004. Punch card voting systems have seen a similar decline, though not as dramatic as paper ballots or lever voting machines. We see an explosion in the use of optically scanned ballots after 1988 and a strong increase in the use of electronic or DRE voting systems from 2000 to 2004. Clearly, after the 2000 presidential election, election officials throughout the nation have moved rapidly to acquire one of the other two voting technologies that are based on some form of electronic technologies—optical scans or true electronic voting devices.

Like punch cards, optical scan ballots involve marking choices on a paper; that paper is then scanned by an electronic device located either in the precinct or at a centralized tabulation location. Thus, as the medium that the voter interacts with is paper-based, an optical scan

TABLE 2.1
Use of Voting Equipment in 1980, 2000, and 2004 Elections

	Percentage of Counties Using Technology		
	1988	2000	2004
Paper ballot	32.7	8.8	10.7
Lever machine	29.3	14.8	7.9
Punch card	25.0	20.9	9.7
Optical scan	5.3	41.2	49.9
Electronic (DRE)	1.9	11.5	19.0
Mixed	5.8	2.8	2.8

Source: These data orginated from Election Data Services, Inc, but have been updated by Delia Grogg and Charles Stewart. See Grigg 2005

voting system faces many of the same problems that paper and punch card voting systems face: mistakes, accessibility, and complexity.

Voters do make mistakes when they use optical scan ballots; they are equipped with a pen and a piece of paper, with nothing standing in their way of failing to vote in a race for which they are eligible (undervoting), for voting more times that are allowed (overvoting), making mistakes with write-in candidates, or in other ways defacing or spoiling the ballot. These mistakes are common with optical scan voting systems, but luckily there are relatively simple and cost-effective solutions to these problems that election administrators have implemented in recent elections.

As noted, a device located in the voting place or in a centralized tabulation center can read the optical scan ballots. The use of tabulation devices in voting places has been argued by many to be a way that voters can check their ballot before casting it; and in many locations, such precinct-based optical scanning appears to lead to fewer mistakes by voters, as measured by residual votes, overvotes, and undervotes (Alvarez, Sinclair, and Wilson 2002). These intuitive reductions in voter errors are likely one of the important reasons that election officials have moved to optical scanning voting devices, when they abandon the outdated punch card, lever machines, and hand-counted paper ballots.

As optical scanned ballots are paper ballots, accessibility becomes a major concern with their use. Voters with vision impairments or a variety of physical handicaps that make it difficult for them to use a pen or marking device can have difficulty casting a ballot without assistance. Generally, these voters may not be able to cast a secret ballot when voting by optical scan, although there are now optical scan products that can "read" the ballot to voters and then mark their choices. However, these technologies use a very different interface than the traditional paper-and-pen optical scan ballot. Voting with assistance also raises concerns over coercion and outright fraud, because individuals providing assistance may be able to alter the ballot without the express knowledge or consent of the actual voter.

Optical scanning ballot systems—at least as they are commonly used today—are difficult to implement in large and complex election jurisdictions.[15] An optical scan ballot has to have all of the choices facing a voter in a particular precinct printed on the ballot, and the ballot must be printed using high-quality paper and printing processes (otherwise the optical readers will have difficulty accurately scanning the ballot for tabulation). This issue is not necessarily significant in small election jurisdictions, where there are few races on the ballot, and where election officials do not need to provide ballots in multiple formats (in the way of either name rotations or multiple languages). Each of these layers of administrative complexity serves to increase the number of types of ballots that must be printed by an election official before an election, thus increasing the cost.

For election jurisdictions that are large and have many required languages (especially in states that use the initiative process), optical scan ballots can be costly and complicated to utilize. For example, in an election with a large number of candidates or referenda, a jurisdiction might have to use either multiple optical scan ballots, or both sides of an optical scan ballot, in order to accommodate all of the candidates and ballot measure races, as well as each required language (Bower 2002).

Because optical scan voting systems are paper-based, they have one attribute that many critics of the current trends in voting technology like: the voter interacts with a paper ballot, which can serve as a lasting record of every vote cast in the jurisdiction. These critics argue that, if disputes arise, these paper ballots can be reexamined and recounted, to check the veracity of earlier tabulations. Of course, anyone who paid attention to any of the recent recounts knows that this is not necessarily the case. For example, the rules governing recounts vary by state and jurisdiction and in states like Virginia, having the paper ballots does not mean that a recount is permissible in all circumstances. Likewise, having a paper ballot and divining voter intent from it are different as well. Voters may not mark their ballot in a way that is easily discernible. In the 2000 election in Palm Beach County, Florida, the "butterfly" ballots contained real votes for Patrick Buchanan; recounting them did not change the outcome. Likewise, people who circled candidate names on an optical scan ballot had to rely on the rules established by local election officials to know if their ballots would be counted. Voter intent rules vary by state, and the implementation of such rules varies across individuals, as we will show in chapter 6.

Also, the "voter-verified paper-trail" argument is predicated on a voter being able to verify that his ballot is recorded as marked. Optical scan counting equipment does not have this capacity. When the voter scans his ballot through the optical scan reader, typically he can verify only that the ballot does not contain overvotes (assuming that the system is programmed to detect overvotes). Scanning to check for errors does not allow the voter to verify how the electronic tabulation software has counted the ballot. This is another example of variation of the rules across voting technologies that affects different voters based on the technology they use to cast ballots. Finally, paper ballots are susceptible to a variety of known threats, both intentional and unintentional: a recent example concerns the use of optical scan ballots in situations of high humidity, where the humidity can cause scanning devices to fail and may even lead to physical alteration of indications of voter intent if ink on the ballot smears or bleeds onto another ballot.[16] However, the presence of a so-called paper trail is one attribute of optical scan voting systems that foes of the broader application of electronic technologies in the election process favor, even though, as implemented and within the context of many state laws, optical-scan voting actually does not provide a paper trail or allow for voter verification. We discuss this debate fully in the next chapter.

COMPUTERS IN POLL SITES?

Many election officials and advocacy groups have pushed for further computerization of poll site voting, but the rationales of this computerization must be considered. One computerized voting device used in poll sites is the optical scan ballot, in which voters mark their choices on paper ballots that are tabulated by a computerized device. Here, although much of the election administration process is dominated by electronic and computerized technology, the voter is still largely interacting with a piece of paper. This paper ballot, in turn, interacts with a computer, which counts the votes by reading—or attempting to read—the paper ballot. The essential issue here is not whether there should be some form of computerized voting, but when and where it should be used in the process.

The move toward computerization of poll site voting is not necessarily aimed at removing the paper from the process, but rather eliminating the problems associated with the voter's interaction with a paper interface. For example, computerized voting devices have been developed and implemented to mitigate or eliminate simple voter errors. For example, both DRE and "ATM-style" voting systems now on the market can block voters from making mistakes like overvotes. The automated-teller-machines style systems can be programmed to tell voters if they have skipped a race and remind them repeatedly that they have failed to cast a vote somewhere in their ballot (minimizing undervotes).

Evidence is accumulating that these features of the newer generations of computerized voting systems are helping to reduce common voter errors. Charles Stewart, a professor at MIT and member of the Caltech/MIT Voting Technology Project, studied the case of the 2000 presidential election in Georgia, which was not a close race. His study showed that the state's voting technologies performed very poorly in 2000: its residual vote rate was 3.5 percent, compared to a national residual vote rate of 1.9 percent, making Georgia second only to Illinois, which had the worst performance of voting technologies in 2000. Stewart has estimated that the Georgia counties using punch card voting systems had a high residual vote of 4.7 percent, closely followed by residual vote rates of 4 percent for lever machines, 3.3 percent for hand-counted paper, and 2.7 percent for optical scanning (Stewart 2004, Table 1).

The poor performance of all of Georgia's voting technologies in the 2000 presidential election, and local concerns that were future elections to be close in that state there could be substantial controversy and litigation, led Georgia's Secretary of State Cathy Cox to acquire and put in place a uniform statewide computerized voting solution in very short order, before the 2002 statewide election cycle. Cox selected the AccuVote touch screen voting system, produced by Diebold. The effect of Georgia's transition to a uniform computerized poll site voting system has been a sweeping reduction in Georgia's residual vote rate. Stewart compared the 2002 statewide races in Georgia relative to the same sort of races that had been held in 1998 and found that in every case the residual vote rate had declined, in some cases substantially. The residual vote rate in the two top-ticket statewide races for governor and attorney general fell by 1.7 and 1.9 percentage points, respectively, for example. In some down-ballot races, such as state school superintendent, some counties experienced reductions in the residual vote rate from more than 25 percent to less than 5 percent from 1998 to 2002. That electronic voting leads to voters' casting a more complete ballot by voting on more down-ballot races is also reflected in studies of elections in Florida.[17] Stewart's analysis (2004, Table 3) of the residual vote rates in Georgia, by counties, documents that these reductions in residual votes were concentrated in rural, heavily African American, and poorer counties in the state. However, Stewart cautions that the Georgia case involves two variables—the transition to a uniform voting technology and strong efforts by the secretary of state's office and Diebold to provide necessary resources to insure that polling

TABLE 2.2
Residual Vote Rates by Year

Machine Type	1988	1992	1996	2000	2004
Punch card	3.5	2.7	3.1	2.6	2.0
Lever machine	1.8	1.7	2.2	2.2	1.1
Paper ballot	2.7	1.9	2.6	2.2	2.2
Optical scan	3.1	3.1	2.4	2.1	1.4
Electronic (DRE)	3.6	3.8	3.3	2.4	1.6
Total	2.9	2.4	2.7	2.3	1.6

Source: Residual vote estimates from Grigg 2005

place workers and voters were educated about how to use the new touch screen voting machines effectively.

More recent evidence from the 2004 election underscores that the newer generations of electronic voting machines appear to help reduce common voter mistakes. This is easily seen in basic data from recent presidential election cycles, using the common metric of residual votes. This means of assessing voting system reliability is based on computing the fraction of ballots cast in a county that have a valid vote counted in the top-of-the-ticket race (here, the presidential election). In Table 2.2 we present residual vote estimates, by voting technologies, since 1988. The national residual vote estimate, after being relatively stable from 1988 to 2000, dropped in 2004 due to the use of new and improved voting technologies improved election administration procedures, and perhaps to greater public awareness of problems with voting systems. Also, evident is a progressive improvement in residual votes for the two technologies of choice in 2004—optical scan and electronic voting systems. For electronic voting systems, which saw relatively high residual votes in the first generation of this technology used in the late 1980s and early 1990s, the drop in residual votes in 2004 is prominent.

A recent analysis by Stewart (2005) documented considerable reductions in residual vote rates, especially in states like Georgia that adopted comprehensive voting system reform, especially the implementation of new voting technologies. Stewart studied a variety of "transition paths" from old to new voting technologies. He found that election jurisdictions that moved from punch card to electronic polling place voting systems had a 1.46 percentage-point reduction in residual vote rate, the greatest for any common transition path in his data. He also found that the transition from optical scan to electronic polling place voting had a relatively large reduction in residual votes, 1.26 percentage points. By contrast, the 154 jurisdictions included in Stewart's data that moved from punch card to optical scanning voting systems saw a 1.12 percentage-point reduction in residual vote rates. As Stewart notes, however, it is uncertain whether these reductions in residual votes are due to technology or to better education and training effort or whether they will be transitory or lasting.

Electronic voting technologies will not necessarily lead to a reduced residual vote rate in all races on the ballot. In the November 2006 election, there was a large undervote in the Thirteenth Congressional District race in Sarasota County, but only for those voters who cast ballots using the Elections Systems and Software (ES&S) touch screen voting systems. In a preliminary analysis of election results, we estimated a 12.92 percent undervote rate (more than 18,000 undervotes from about 140,000 ballots cast) among all ballots cast.[18] The undervote rate was 13.9 percent for election day electronic voters and 17.6 percent for early electronic voters. By contrast, absentee voters who used an optical scan ballot had an undervote rate of 2.5 percent. In neighboring Manatee County, where voters cast ballots in the same congressional race but used an optical scan ballot, the undervote rate was 2.4 percent. The primary culprit for this high undervote rate in Sarasota County in the congressional race would seem to be bad ballot design. The congressional race was located at the top of the screen, above a banner that separated state races from federal races, and voters could easily have overlooked the congressional race because the banner drew their vision down.[19] This example shows that any voting system, including an electronic voting system, can be subject to ballot design problems.

Another justification for the movement toward the use of electronic voting machines in poll sites, especially the newer touch screen–style electronic voting systems, is accessibility. These voting devices are strongly favored by many advocates for disabled or physically handicapped voters, who claim they allow these voters to cast secret and coercion-free ballots. Touch screen–style electronic voting systems can be (and are often) equipped with an audio system; a blind or visually impaired voter can use headphones, to navigate the electronic voting device without additional assistance. Physically handicapped voters who might otherwise have trouble using pens or other objects to make marks on paper ballots can sometimes better navigate an electronic voting system equipped with either a touch screen or wheel-like scrolling device; other means are available for those unable to touch or directly manipulate the electronic voting device.

Also, electronic polling place voting systems can often solve administrative problems in larger and more complicated election jurisdictions. The newer electronic polling place voting devices, which essentially use a computer to run some sort of graphical interface for the voter, can be programmed to provide that interface in a wide variety of ways.

Electronic systems can be programmed to display a ballot in languages other than English, thus helping resolve an increasingly common problem for election administrators. Some systems, especially the newer generations, can easily display long ballots because they are typically set up with an interface that permits the voter to move sequentially through the ballot, with each screen representing a different race or decision. The only real limitations on ballot length are the capacity constraints of these computerized devices.

The newer electronic voting machines can be programmed so that voters can receive a variety of ballot styles. Election officials like this feature because the newer electronic voting machines can now be used in early voting and voting "supercenter" situations. For example, the election official can simply place a small number of electronic voting stations in a variety of public places before the election, and any registered voter from the jurisdiction could receive the correct ballot when they go to vote—the computer technology inside the voting system can store and retrieve any valid ballot type for the entire jurisdiction, which can easily number into the hundreds, perhaps the thousands, in very large jurisdictions with very complicated ballots.

This feature may prove valuable in the future, as we learn more about ballot designs and appropriate presentation of choices to voters. For example, there is a growing literature on how the placement of a candidate's name on the ballot may provide some slight advantage or disadvantage, which normally may not be a concern but which in a very close race or low-information and down-ballot contests could influence election outcomes (Alvarez, Sinclair, and Hasen 2006; Ho and Imai 2004; Koppell and Steen 2004; Miller and Krosnick 1998). If research in this area indicates that candidates at certain focal points on the ballot receive subtle vote advantages, it would be possible to program electronic voting systems in the future to randomize the order of candidate names for each voter.[20]

Usability studies conducted by a research group based at the University of Maryland on both electronic voting machines and paper ballot voting technologies have also identified interesting findings.[21] Not surprisingly, there are differences in how voters rate the usability of different electronic and paper voting systems, based on the features they provide and the ballot style requirements that a state utilizes. For example, voters tend to make more errors in voting when a straight-party option is provided, compared to a ballot without this option. One of more interesting and surprising findings relates to write-in voting. When voters were asked to write-in a candidate's name on electronic and paper systems, the voters could complete the task faster on a paper ballot. However, the voters also made many more errors on the paper ballot including failing to check the box indicating a vote for the write-in candidate that is required under state law in many states. Without the box being checked, the vote would not count. Such studies illustrate the complex interactions that exist between voting technologies and both state legal requirements for ballot design and ballot counting requirements.

Also, current research on voting system usability may in the future lead to innovative new ways to use electronic voting technologies to better present ballots to voters. A primary example of the exciting and innovative research now being done on electronic voter interfaces is the Low Error Voting Interface (LEVI) being developed by MIT Media Laboratory professor Ted Selker (who also is codirector of the Caltech/MIT Voting Technology Project). The LEVI interface employs a number of ballot interface improvements that might help voters navigate future ballots and electronic voting devices. For example, the LEVI system uses tabs on the side of the ballot, representing all races on the ballot—texture and color identify for a voter when a selection has been made. The LEVI system also uses a "fisheye" approach when there are a lot of races on the ballot. This involves decreasing the resolution and font size of the text when voters move further from the current selection; when they move within the ballot, the resolution and font size increase and size and resolution on those parts of the ballot they are moving toward. The LEVI system is currently undergoing experimental testing and further development, but initial tests indicate it helps reduce voter errors (Selker et al. 2005).

THE PROS OF ELECTRONIC-VOTING

This chapter has provided a history of the evolution of voting technologies in the United States and has reviewed the basic rationales that have been advanced for the continued adoption of electronic technologies in the American election process. Evidence from recent elections suggests that increasing the use of electronic technology in the election process has the potential to produce a more accurate, accessible, and easily administered voting system.

We can also see how the movement to electronic voting represents a decision to address concerns about the risks associated with previous methods of voting. Voice voting and party-centered paper ballots were effective means of voting—voice voting even encouraged both voter verification and auditability—but they entailed other risks, such as coercion, violence, and vote buying. The movement to the secret ballot was designed to address these risks, but it introduced new ones, such as ballot fraud. Lever machines and then electronic voting represent an effort to address concerns about these new threats, but they too have brought new concerns to the fore. One issue with the current debate over electronic voting is that there are ongoing efforts—such as the voter-verification debate—that threaten to create a bifurcated set of election laws that do not require one technology (paper ballots) to meet the same standards as another technology (electronic voting).

As we consider in the next chapter, electronic voting technologies raise many concerns. These focus largely on the security of electronic voting— whether they help achieve a more secure and less fraud-prone election process, or whether they simply open the door for new (and possibly large-scale) attacks on our electoral process. Others are less concerned about the security of electronic voting and are more worried about the fact that the reliance on increasing levels of technology in the election process raises the potential for more glitches and unforeseen errors in the balloting and vote tabulation. These concerns have merit, as we also discuss in the next chapter. The critical issue is to achieve positive outcomes through electronic voting systems, while minimizing or eliminating the concerns about glitches and outright fraud.

Chapter 3

CRITICISMS OF ELECTRONIC VOTING

In this chapter, we discuss the recent criticisms of electronic voting. To highlight the fears of electronic voting, we begin with a fictionalized story that illustrates the concerns of some critics.

It is early in the morning on November 5, 2008. With no incumbent in the race, both the Democratic and Republican presidential candidates have been running neck and neck in the polls all year. As the early election results start rolling in, the election is again extremely close and the pundits start to rave about how close the Electoral College vote will be. By two thirty on the East Coast, we are down to the votes in one state, and the Electoral College votes of this state hang on the difference of only a few hundred ballots. Whoever wins this state will win the presidency. All throughout election day, there were reports of widespread election anomalies across the country—but there were many reports of problems coming from this single pivotal state. It is clear something is wrong because this state uses electronic voting equipment in the entire state without paper audit trails—although different counties use equipment from different vendors—and e-voting should be producing vote totals much quicker. In fact, Georgia, which also uses electronic voting equipment, reported complete totals for the entire state in less than two hours after the polls closed.

The state's secretary of state comes onto the television at three o' clock in the morning and reads an incredible statement:

> Ladies and Gentlemen: For the last five hours, we have been attempting to collect vote totals from all of our counties. In our ten smallest counties, we were able to collect vote totals quickly, without problem. However, I am here to report that every voting machine in our largest county—which has 500,000 registered voters and approximately 400,000 cast votes in this election—is blank. The machines did not register a single vote for any race. County election officials and the voting machine vendor are currently attempting to determine if the ballots are in either of the two redundant data storage devices within the voting machines. We have examined fifty machines from across the county, and none of the machines show a single ballot in any memory registry.
>
> This problem is obviously disastrous. Unfortunately, it is not the only problem we have. In our two medium-sized counties we have complete election results for all races. However, the results for the presidential race are highly

anomalous in both counties. In the first county, 80 percent of the votes are for the Democratic candidate, but 65 percent of voters voted for a Republican in each of the next three races. In the second county, we have 12,453 more votes than we have voters who cast ballots, and again, the votes are anomalously skewed toward the Democratic candidate. In both 2000 and 2004, these counties voted 60 percent for George W. Bush. Election officials in these counties are working with the voting machine vendors to determine if the vote totals are anomalous.

This hypothetical case combines into a single example many of the fears that critics of electronic voting have about this technology: machines that lose ballots, anomalous results, an inability to correct the problem, and the potential of obvious fraud. Even this hypothetical story, however, fails to capture fully the concerns of critics of electronic voting, because in this scenario the snafus and possible fraud were so significant that they were easily noticed by election officials and observers. Many of the critics of electronic voting are distressed about deeper problems than these—for example, situations where someone could hack an election and switch just enough votes to change the outcome, but do so in a way where no one would even be aware the fraud had occurred. Someone would have stolen an election, and no one would know.

ELECTRONIC VOTING IN THE RISK SOCIETY

The critiques of electronic voting are typically presented as an engineering problem without consideration of how other events and phenomena in society affect our views of electronic voting. Although the potential problems with electronic voting are well elucidated by the critics, they are rarely placed within a sociopolitical context or framework, as is done in similar debates in other policy areas. We put the debate over electronic voting within this political framework in chapter 4, but we focus on how electronic voting fits within a broader sociological construct about risk.

Electronic voting can be seen as falling within the rubric of what Ulrich Beck refers to as the "risk society."[1] In the risk society the world is a giant laboratory with one exception: we conduct experiments without the benefit of the controls we have in modern scientific laboratories. The world itself serves as the laboratory, and any benefits or failures can affect broad swaths of the population. This is a world where our efforts to control risk through modernization—especially through the use of technology—have the potential to create additional risks in our lives. We attempt to treat infectious diseases aggressively with antibiotics and can end up creating drug-resistant superbacteria. We have "fast-track" approval for novel medicines that later are found to have adverse and unforeseen consequences for millions of patients. We create genetically modified foods and open the possibility of uncontrollably altering our ecosystem and harming mankind.

Risk used to come solely from nature. Today, we live in a world of manufactured risk, where the innovations designed to improve our lives sometimes instead lead to vast and expensive future problems. With natural risks, we attempt to make the "unforeseeable foreseeable" through the use of a calculus of risk, like an insurance actuary (Beck 1998). We know that there is a risk of fire to our homes, so we purchase home insurance. We purchase health and life insurance to cover ourselves in the case of disease, accident, or death. These purchases occur in a marketplace where actuaries calculate the risk of an incident befalling us and, using a wealth of data, set a price for our insurance. In the case of risks that are very large, the price is either exorbitant, a governmental entity provides some degree of insurance against risk, or there is no market for insurance. In the contemporary context—especially in the wake of Hurricane Rita's near-destruction of New Orleans—the issues of risk and insuring against this risk have become much more a part of the public debate than in recent memory.

In the risk society, many risk potentials are simply unknown and impossible to predict. Thus, the world is an unpredictable place where "sciences are operating in terms of probabilities, which do not exclude the worst case"; the worst case scenario—or what Beck refers to as the "worst imaginable incident" (WII)—can mean the destruction of mankind through a biotechnological, nuclear, or environmental disaster(1998, 13, 53). Once a WII occurs, it may be impossible roll back the clock and restore the status quo. The risk of such an event occurring is complicated by the fact that, in an array of high-technology and complex scientific areas, the risks associated with these new creations can be assessed only *after* they are developed and implemented. We can assess the risk only when the threat exists in the real world and cannot be recontained.

Paradoxically, our high level of knowledge and the vast amount of information that we generate daily expand the risk problem. With new information, things that were previously risk-free—in our perception—become quite risky. In the United States, recent issues associated with COX-2 selective inhibitors illustrate this point. In the eyes of the public there was every indication that these drugs were safe, until one of the manufacturers pulled its version of the drug after learning that such drugs potentially carried inordinate risks for users. Other manufacturers, however, continued to market the drugs, leading to public debate about the risks of COX-2 inhibitors, especially by researchers and experts from both sides. The media coverage was intense but often lacked basic information that the public needed to understand the debate. Not surprisingly, the debate left the public confused, unsure what it could do other than not purchase some drugs deemed previously safe.[2] This debate was made even murkier because doctors who study drug risk viewed this debate much differently than did doctors who prescribe the medication. Much like the debate over electronic voting, the practitioners and the researchers often came to different conclusions about risk when looking at the exact same evidence.

For Europeans, the problem of the risk society was brought to the fore in the 1990s by the outbreak of bovine spongiform encephalopathy (BSE) in cattle and variant Creutzfeldt-Jakob disease (vCJD) in humans,[3] an outbreak that threatened the European food supply. In the case of BSE, there is evidence that experts in the United Kingdom failed to respond quickly to the threat, even when presented with a wealth of information; instead, they denied there was a problem until it was too late.[4] The potential risk of a BSE outbreak was known well before a large-scale outbreak occurred. However, government experts UK and the policy makers to whom they report did not act in time to avert a crisis requiring the destruction of more than a million head of cattle in the United Kingdom and which led to the spread of both BSE and vCJD on the continent.

Of course, not all proponents of the risk society think that the world is actually becoming riskier. As one scholar noted, "it is a society increasingly preoccupied with the future (and also with safety) that generates this notion of risk."[5] A variety of factors—increased access to information, the rapid distribution of information using new technologies like the Internet, the generation of new scientific knowledge, and the media's focus on conflict—all serve to raise fears and generate the conditions in which the risk society exists (Mythen 2004, 140–142). We are ever more aware of the potential risks around us; the vast improvement in computing power allows us to measure risks and analyze data in ways that were not possible even ten years ago. These threats are often made more frightening because the extent of the risks is unknown. We cannot be sure if what seems safe today will be safe tomorrow, or if we have the capacity to overcome a catastrophic event, should it occur. This view of risk also often presumes that new risks are more dangerous than old risks. Recent food scares related to *Escherichia coli* bacteria illustrate the problem of underestimating old risks; according to the Centers for Disease Control and Prevention, "foodborne diseases cause approximately 76 million illnesses, 325,000 hospitalizations, and 5,000 deaths in the United States each year."[6] Sunstein (2005) argues that this focus on overestimating new risks and underestimating existing risks is a problem that is common in risk analysis, and it is not surprising that such risk analysis occurs rarely in relation to electronic voting.

Electronic voting fits into this understanding of the risk society because it is technological in nature and because it too is potentially susceptible to a catastrophic event. The hypothetical at the beginning of the chapter illustrates the potential scope of such a calamity. An election could occur in which the actual outcome was never known. If this happened, the legitimacy of a democratic election could be thrown into question and force a political solution to the democratic process of selecting members of Congress or a president. After a presidential runoff election in the Ukraine, on November 21, 2004, allegations of fraud, led to massive public protests by supporters of presidential candidate Viktor Yushchenko and the Ukrainian Supreme Court's invalidation of the election results. A revote on December 26, 2004, led to a reversal of the initial runoff results and to the establishment of Yushchenko as president of the Ukraine.[7] Unfortunately, such a revote might not be possible in federal elections in the United States because the Constitution sets the sole date for when such an election is to occur.[8] The Ukrainian example is also helpful because it is an example of fraud using traditional paper ballots, underscoring that there are important risks that exist with the use of the baseline technology of elections—the paper ballot.

Electronic Voting: The Critique

In the aftermath of the 2000 election, many people viewed electronic voting as a potential panacea to solving the problems that had occurred in Florida. There, the full array of flaws with paper ballots had been brought to the fore, problems that were well known among election officials but not the general public. However, reports by the Caltech/MIT Voting Technology Project (VTP) issued in early 2001 were among the first to focus public attention on the potential issues associated with electronic voting.[9] These reports found that, of the four primary voting technologies in use—electronic voting equipment, lever machines, optical scan ballots, and punch cards—electronic voting technologies had relatively high residual vote rates. This criticism was not well received in the election community, where many officials thought that the performance of older electronic voting technologies still in use was masking the higher-quality performance of a new generation of electronic voting machines.

Interestingly, the debate over electronic voting machines that occurred in 2001 centered on the ability of these electronic voting technologies to record votes more accurately than other voting technologies. Although some critics raised other concerns about electronic voting at this time, these concerns were not presented as being widespread, and there was not a strong base of peer-reviewed academic research on this subject.[10]

In July 2001 the Caltech/MIT Voting Technology Project summarized the security concerns of electronic voting as:

- the loss of openness (we can no longer see the count)
- the loss of the ability for many people to be involved in the process (we cannot observe its accuracy at various points in the process)
- the loss of the separation of privilege (we vest all control into a single entity—the voting machine or vendor)
- the lack of redundancy and true auditability (we may not have sufficient audit trails)
- the loss of public control (we rely too heavily on various players in the process)

This critique mirrors the concerns that have been raised by international organizations who study elections and have noted that there is a strong need for elections to be transparent and for audits to be used when the vote counting process is not transparent.[11] It was not until after the 2002 general elections, however, that concerns about the security of electronic voting became widespread and studies examining some of the actual security flaws that existed in electronic voting technologies were disseminated.

Defining the Problem

Michael Shamos (2004), an internationally recognized expert on electronic voting, has written an analysis of the criticisms of electronic voting and the need for a voter-verified paper audit trail. He lists some of the arguments that have been made against electronic voting:

- Electronic "voting machines are 'black boxes,' whose workings are opaque to the public and whose feedback to the voter is generated by the black boxes themselves. Therefore, whether or not they are operating properly cannot be independently verified."
- "No amount of code auditing can ever detect malicious or even innocently erroneous software. [Relatedly,] no feasible test plan can ever exercise every possible combination of inputs to the machine or exercise every one of its logic paths."
- Hackers have proved adept at hacking numerous Web sites, illustrating the vulnerability of electronic media.
- Companies that make electronic voting machines may be run by individuals with partisan political interests, and those individuals may order that their voting machines be programmed to record and count votes in a way that reflects this bias.

- There have been many failures involving electronic voting machines in elections around the country that resulted in permanently lost votes and possibly changed election outcomes, illustrating the fragility of this voting technology.

In short, this critique suggests that electronic voting machines may be vulnerable to attacks including insider attacks, that could be difficult not only to prevent but also to detect. Many critics argue that such problems could overwhelm any potential benefits that accrue from the use of electronic voting equipment.

The black-box problem is the initiation point for discussion of other vulnerabilities with electronic voting machines.[12] In "black-box" systems, it is possible to know what goes into a system, and to know what comes out of a system, but not know what happened when the input was being processing by the system. Black-box problems are not limited to computers or technologies generally. Anyone who has ever picked up a newspaper after Congress has adopted an omnibus piece of legislation recognizes that lawmaking too can also suffer from the black-box problem. Consider the cases where various provisions have been inserted anonymously and at the last minute into a piece of legislation and not noticed until well after the bill has been passed by both chambers.[13] We know that something happened—we can see it in the outcome—but we do not know exactly the mechanism by which the change occurred. Thus, black-box problems are common in our lives, although the term is typically used mainly in a context of high technology.

All voting technologies used in public elections with secret ballots have certain properties that exacerbate the black-box problem. In most cases it is not possible for voters to determine that their vote was counted (not to mention counted accurately) because we generally cast secret ballots.[14] Once a ballot is placed into a ballot box, in all but a very few cases there is no way to trace a ballot back to a certain voter. At this point, the voter loses track of that specific ballot and can only "see" the vote in an aggregated vote total for all candidates after the election is complete. The voter cannot determine, however, if the specific ballot cast was included in the vote total. All the voter can do is trust that the ballot placed in the box is included in the final tabulation, unaltered. If concerned, a voter can monitor the ballot box in the polling place, she can watch it be transported to the tabulation location, and follow the ballot box as it is unsealed and the ballots examined and tabulated. But even with all of this scrutiny it is impossibleto follow an individual ballot from casting to counting, with certainty.[15]

In an electronic voting scenario, the same problem arises: the voter makes a series of selections on a screen, confirms the selections, and then "casts" a ballot into an electronic ballot box. In actuality, however, all the voter can do is see what the computer displays on the screen. The voter has to trust that what the computer displays to the voter are the vote choices that the computer has recorded into its digital memory—its electronic ballot box. This system creates what some have referred to as a "man behind the curtain" problem; you are giving your vote to a software system and hoping that it puts your ballot in the ballot box without making changes to your choices. Just as with paper-based voting, the voter who votes on an electronic voting machine cannot follow an individual ballot from casting to counting and know with certainty that the ballot was counted accurately.

Interestingly, most computer systems have this same problem. When we deposit money into an ATM, we do so trusting that the bank will remove money from the deposit bin in the ATM and place the money in the correct bank account. We trust that when we purchase gas and use our credit card, that the purchase will be accurately recorded against our account. When we send an email message, we trust that the innards of our computer will parse the words we type into smaller packets correctly, that those packets will be sent error-free across the country (or world), and that they will be reconstructed again correctly on the receiver's end.

The difference between electronic voting and most other electronic transactions is that the latter processes produce receipts (and electronic records) that the customer can later use to prove that a specific transaction occurred because these records have identifying information associated with them (account numbers, transaction numbers, and even the name and address of the individual undertaking the transaction). In most electronic transactions, there are also redundancies. For example, in the electronic banking process or when you fill up your car with gas, a video camera may record your transaction as well. Also, unlike electronic voting, most electronic transactions make no effort to strip the identity of the individual from the transaction. Every action of a banking client is recorded in the bank's records. Every use of our credit cards is recorded to our account record. When you make a deposit, withdrawal, or purchase, your account number and other related information are attached to the amount of the transaction. At the end of the day, the bank can fully audit its accounts and determine if there are any errors. Furthermore, if you claim that there was a mistake, your transactions can be reviewed to determine what the error might be. And there exists a strong legal and regulatory framework so that fraud or mistakes in electronic transactions can be resolved.

Under federal law, for example, after reporting the loss or theft of a credit card, a consumer has a maximum liability of only fifty dollars for use of that card.[16] In voting, due to our current convention regarding ballot secrecy, identity is stripped from a vote the moment a ballot is cast. There is no way in the current election administration process to construct a complete audit that links every vote cast to every voter so that the voter could verify the accuracy of the cast vote record after the election is complete.

Although e-banking is a black-box activity, it is one where the client is in a position to use other documents—receipts, canceled checks, the ATM surveillance tapes—to correct any error. In the current voting situation, with both electronic and paper voting technologies, there is no way ascertain that every voter's ballot was recorded precisely as intended, because of the break between the voted ballot and the identity of each individual voter. The election official and the voter alike are at the mercy of the data that come out of the machine at the end of the day and are not in a position to independently audit whether these data reflect the actual ballots cast by the voter. In the end, anonymity in voting makes the problems associated with solving black-box problems much more difficult than in other areas.

THE MANCHURIAN CANDIDATE

At the opening of this chapter, we presented a scenario where an election is severely affected by a series of incidents occurring on electronic voting platforms. This example presents the potential outcome of a breach of an e-voting system. By considering a specific security exploit—the so-called Manchurian exploit—we can see how the principles we associate today with voting, especially the secret ballot, leave voting technologies vulnerable to security failures that are potentially undetectable. In the movie *The Manchurian Candidate* (1962), the Chinese and the Soviet militaries work together to capture an American military squadron and then brainwash the members.[17] One member—Raymond Shaw—is brainwashed so that when he hears the phrase, "Why don't you pass the time with a game of solitaire?" he enters a trance. As Raymond plays the game, whenever he sees the queen of diamonds, he does whatever he is told, be it kill a man or jump into a lake in Central Park. What makes the *Manchurian Candidate* so frightening are Raymond's external attributes: he is completely normal when not under this induced hypnotic trance, retaining his previous personality. In addition, his credentials include being a Congressional Medal of Honor winner (which he wins based on the testimony of his brainwashed comrades).

Although this is a work of fiction, two computer scientists, Michael Shamos and Ron Rivest, have suggested that something similar could occur in the realm of electronic voting. The *Manchurian Candidate* voting machine would be brainwashed by its controller—a computer programmer or technician who loads the software onto the machine before it is deployed—so that when a voting machine is told the secret phrase, it will do what it is told.

How would this work? The secret phrase could work in a variety of ways. For example, a voter might type the phase "Manchurian Candidate" as the write-in candidate for president or vote for every other race, starting with the first race, and skip each race in between. Or the voter might simply touch each corner of the voting machine's screen in some secret pattern, or insert a special smart card into the voting machine's card reading device. Once a voter has entered the secret code, the attacker could be given access to vital components of that voting machine. The secret code could activate a prepositioned application that the attacker could use, or it could allow the attacker the opportunity to add or change the software that is operating the voting machine.

With this control, Rivest has outlined one nefarious way the machine might be programmed to behave. Specifically, the voter with control enters a complete state of ballot choices, and the machine records these choices as the "ideal ballot." Then, as each subsequent voter casts a ballot, some percentage of ballots cast for "nonideal candidates"—say 5 percent of those votes—are presented to the voter correctly on the summary screen but recorded in the machine memory with the "ideal" ballot choice. Of course, this attack could operate in many ways, where the attacker could just force the voting device not to record a vote that is cast for a candidate the attacker opposes or could entirely disable the voting device so that it records no votes at all when the voting period ends and election officials are attempting to conduct an initial tabulation of votes.

The *Manchurian Candidate* attack, as outlined by Rivest, is quite elegant because it allows the attacker to thwart many existing efforts to detect attacks on electronic voting machines. Also, it does not require the attacker to specify which parties or candidates should benefit from the attack ahead of time and, unless detected, is reusable from election to election. There are weaknesses in this particular attack, though. One important drawback is that it is an attack on a single machine that might, at most, collect 300 votes over the course of the day, and switching 5 percent of the votes on the machine (15 votes) results in only a 30-vote swing between candidates. Another important weakness of this attack on an electronic voting system is time, as polling place workers and other voters would be likely to notice if the attacker took an inordinate time in the voting booth or was observed doing odd things with the electronic voting device. A last weakness with this attack is that it would require the prepositioning of some application or module in the electronic voting machine's operating system, and the odds that such an anomalous application or module would be detected in system testing may be greater the larger the size and utility of the nefarious code.

Much bigger swings in these vote totals could occur if the programmer could create a *Manchurian Candidate* machine at a higher level in the election process. If a magic phrase existed at the county level that would affect both the central tabulation of results and the code on the voting machine memory cards, then a single person would be in the position to switch 5 percent of more than a million ballots in some jurisdictions. Or if a secret code existed in the software application that a state uses to aggregate the election returns from all of the counties into a statewide total, it could influence the outcome of a national election. If all election officials in a state were using the same administrative software application, the attacker could change data at both state and local levels, and such an attack might be very difficult to detect in the immediate aftermath of an election.

THE DIEBOLD EXPERIENCE: FINDING FLAWS IN ELECTRONIC VOTING CODE

The debate about the security of electronic voting became more concrete in July 2003 when computer scientists analyzed the source operating code for a version of the Diebold touch screen electronic voting equipment.[18] The code was obtained from the Internet, where it was downloaded from an improperly secured ftp site operated by Diebold.[19] According to the comments within the code, the code was generated in April 2002, and it is unclear whether it is a final source code file. The analysis considered eleven possible attacks on the system and then considered whether the code was written in a way as to minimize the likelihood of these attacks occurring. The attacks analyzed included the ability to vote multiple times on a forced smartcard and to create, delete, or modify votes on the system. The potential perpetrators considered ranged from a voter (who had access to a forced smartcard) to the operating system or voting device developers.

Not surprisingly, the analysis found that the developers of the software and the voting device were in a position to write the code in a way that would allow a full menu of attacks—with the exception of voting multiple times on a forced smartcard—to be perpetrated. This assumed, of course, that the developers' schemes are identified neither during the certification process, during logic and accuracy testing, nor during use of the voting machines. It also found that any data that traveled over an insecure network could be compromised, which would allow for six of the eleven attacks to occur. Likewise, a poll worker able to access the storage media on each electronic voting machine could do everything from changing the votes on the machine to linking voters with their votes. In general, the analysis found that the code was not well written, did not employ effective cryptography, and did not implement basic security protocols, such as the use of unique "personal identification numbers" (PIN) on various smartcard devices. The conclusion was that "this voting system is unsuitable for use in a general election. Any paperless electronic voting system might suffer similar flaws, despite any 'certification' it could have otherwise received." (Kohno et al. 2004, 1).

After the preliminary report was first issued by a Johns Hopkins University and Rice University team studying the Diebold source code, the governor of Maryland, Robert Erlich, hired a professional computer security firm, Science Applications International Corporation (SAIC), to examine the Diebold system and its implementation in the context of Maryland elections. In addition, RABA Technologies was hired by the Maryland Department of Legislative Services—an arm of the Maryland General Assembly—to evaluate both the Johns Hopkins report and the SAIC report. These reports were intended to provide policy makers with guidance on the question of what to do with the Maryland Diebold system in order to overcome any deficiencies it might have, as identified in the Johns Hopkins/Rice report that had received considerable public attention.

The SAIC report provides a nuanced set of findings:

> In general, SAIC made many of the same observations [as were made in the Johns Hopkins report], *when considering only the source code*. While many of the statements made by Mr. [Avi] Rubin [of Johns Hopkins University] were technically correct, it is clear that Mr. Rubin did not have a complete understanding of the State of Maryland's implementation of the AccuVote-TS voting system, and the election process controls or environment. It must be noted that Mr. Rubin states this fact several times in his report and he further identifies the assumptions that he used to reach his conclusions. The State of Maryland procedural controls and general voting environment reduce or eliminate many of the vulnerabilities identified in the Rubin report. However, these controls, while sufficient to help mitigate the weaknesses identified in the July 23 report, do not, in many cases meet the standard of best practice or the State of Maryland Security Policy.[20]

In short, the SAIC analysis confirmed many of the findings in the Johns Hopkins/Rice report regarding deficiencies in the code quality and the security vulnerabilities that existed in the Diebold AccuVote-TS system.

However, the SAIC report concluded that these problems are somewhat less dramatic within the context of an actual election, when voting systems are implemented within a specific regulatory regime. SAIC identified seventeen mitigation strategies that, if implemented, would address the high-risk issues that were identified in the Johns Hopkins/Rice report and make the voting system work effectively in an elections setting. If these strategies were not implemented, the system would remain at a high risk of being compromised during an election.

The RABA report, likewise, examined both the Johns Hopkins/Rice and SAIC reports. RABA also conducted and reported the results of a "Red Team" exercise, where computer security experts tested the security of all aspects of the system.[21] As was the case with both the previous reports, the Red Team found significant vulnerabilities with the system. And as was the case with the SAIC report, the RABA report makes a specific set of recommendations regarding how the vulnerabilities identified in the report could be mitigated.

In both the RABA and SAIC reports, the authors note that no electronic system is without vulnerabilities. However, the RABA report explicitly asks the question regarding what level of security should be required before a system is implemented and suggests that the Johns Hopkins/Rice report sets the bar much too high. As we noted previously, however, in this concept of a risk society, even small likelihood events have to be treated seriously, because, although small, these events can happen. SAIC noted specifically that any system failure could have a dramatic affect on public confidence in elections and in Maryland's election governance. In principle, a failure could undermine our democracy by undermining the confidence that key stakeholders (like voters, politicians, and the media) have in the legitimacy of the process.

The Diebold touch screen voting system has been subject to additional study by computer scientists at Princeton University.[22] Unlike the study done by the Johns Hopkins University and Rice University researchers, the Princeton team studied an actual Diebold voting machine, not just the software from a voting machine. The key new result presented by this team was to note that a virus could be spread over time to other voting machines if such a virus was installed in the memory card of the voting machine. The propagation of such a virus would occur over time and require that the infection not be detected by any system testing. As research by Election Science Institute (ESI) in 2006 showed, attacks involving electronic voting system memory cards might be difficult to detect in some jurisdictions with relatively lax security procedures.

ESI conducted a study of the primary election in Cuyahoga County, Ohio, in May 2006.[23] Cuyahoga County's 2006 primary was an interesting case—the county was implementing the Diebold electronic voting machine, with a voter-verified paper audit trail, for its first large-scale election. The study of this electronic voting system roll-out was unique in its comprehensiveness; data were collected about voter attitudes, poll worker attitudes, precinct incidents reported by poll workers, and machine functionality, through a postelection machine audit. From these data, especially the incident reports, we can identify key problems that occur with electronic voting machines in an actual election.

We created two figures for the ESI report showing the problems in precincts and the overall problems in the election (Figures 3.1 and 3.2). A sizable percentage of precincts had some voting machine problem. In addition, a large number of precincts had some ancillary problem with their voting machines, including problems with the seals that secure the machines, ballot printer failures, and encoder or access card failures. Many of these machine problems are distressing, from a security per-spective, as they indicate places where the security of the electronic vot-ing device could be compromised. The data indicated that 4.2 percent of all reported incidents were related to broken seals on machines and paper canisters or to a shortage of seals to secure all of the voting ma-chines at the beginning of the election. Approximately 4 percent of all reported incidents involved the voting system printer, typically reported as a failure or paper jam. And another 4 percent of reported incidents in-volved access cards, card encoders, and memory cards: compromising the voter access cards, their encoders, or the memory cards could allow someone to manipulate an election outcome.

Although much of the discussion of the ESI study focused on the prob-lems with the voting machines and with the ESI team's difficulty in au-diting the voting machine results, other problems were associated with the May election in Cuyahoga County. The failure of poll workers to show up on election day may have exacerbated some voting machine problems. In addition, the effectiveness and level of training of the poll workers who did report to their precincts may also have had a negative impact on security and functionality of voting technologies.

Security tests have been conducted on other voting machines and these tests have illustrated that all electronic voting machines do not suffer from the same security weaknesses. For example, a study conducted for Alameda County in California found that the Sequoia AVC Edge elec-tronic voting machine did not suffer from the same security problems that have been identified with the Diebold electronic voting machines.[24] In fact, the report contains a table that makes a side-by-side identification of potential security vulnerabilities and identifies whether the Diebold and Sequoia systems suffer from these potential vulnerabilities. The Table shows that the Sequoia system is susceptible to 1 of 12 vulnerabilities but the Diebold system is susceptible to 10 of the 12 vulnerabilities. This study

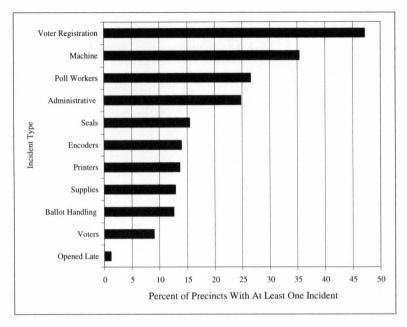

FIGURE 3.1 Cuyahoga County, OH, Precincts with at Least One Poll Worker-Reported Incident, by Incident Type
Source: Data collected by authors, Jonathan Katz, and Rod Kiewiet

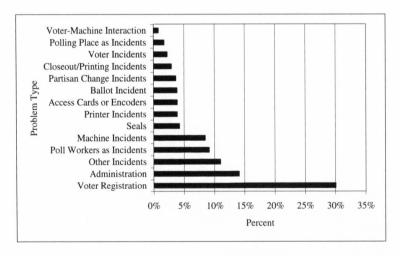

FIGURE 3.2 Cuyahoga County, OH 2006, Frequency of Poll Worker-Reported Incidents
Source: Data collected by authors, Jonathan Katz, and Rod Kiewiet

also suggests that generic discussions about electronic voting fail to take into consideration variations that exist across electronic voting systems.

REGULATORY MISMATCHES

The critiques of electronic voting outlined by Shamos and various computer science research teams and the *Manchurian Candidate* attack discussed by Shamos and Rivest show not only the potential vulnerabilities of electronic voting systems but also the regulatory mismatch that exists in the area of voting systems. Federal voting standards have been adopted in 1990, 2002, and 2005. These standards are all voluntary; states are not required to purchase and use voting equipment that meets these standards in federal elections, unless they use federal HAVA funds for the purchase.

What is most problematic, however, is that even if a state does purchase certified voting equipment, this is no guarantee of the effectiveness or quality of the system. The limitations of the voting system standards and any certification process based on them are discussed in a report by the Miami-Dade County inspector general, who when investigating the performance of an electronic voting system purchased by the County, wrote:

> The FVSS [Federal Voting System Standards] . . . sets forth the minimum standards of voting systems and establishes testing and certification procedures to determine if the required minimum standards have been met. . . . But what the FVSS does not evaluate is the ease and operation of the voting system. . . . Simply put—certification means that the product (the voting system) may be used in the State. . . . Certification is by no means a product endorsement; it has no bearing on whether it is a good product. Certification also has no bearing on whether the product will perform to the needs or expectations of any county, small or big. It does not address whether a system is "state of the art technology."[25]

In short, the voting standards provide a floor—what some view as being a very low floor—for determining if a system meets the minimal level of quality to be used in voting.

Moreover, the 1990 and 2002 standards both had few requirements regarding testing the security of electronic voting equipment. However, critics of e-voting argue that the quality or quantity of the security testing is irrelevant. Given the complexity of the code used in a typical e-voting system, the ability of individuals to find security vulnerabilities, and the weaknesses of the existing voting systems standards, there is a chance that security exploits or other system glitches will not be found.

It is thus possible that exploits are now lurking in electronic voting systems, waiting to be activated. The critics' concern about the theoretical inability to find all errors in computer code means that the certification process, as related to security, is irrelevant. Even assuming that the software code that was certified by an independent testing agency could be completely validated as being free of any malicious computer code, it is difficult to verify with certainty that this same code set is the code that has been loaded into the voting machines for use on election day. To illustrate the scale of the problem, consider the state of Georgia, which uses a single electronic voting system. There are 22,000 voting machines used in the state, which expands the problem of examining the code used in elections exponentially. Additionally, the software code in an electronic voting system has to be updated as election laws and local needs change, creating more opportunities for problems to be inserted into the system. Georgia does use important security procedures, like hash code testing, to ensure that each voting system has the original certified software. However, this process is time consuming.

The problem of regulating the software code used in electronic voting systems brings us back to the underlying theme of this chapter: there is a mismatch between the problems we face in the modern risk society and the regulatory framework we have in place to address these problems. Largely, we are using a nineteenth-century industrial regulatory framework to attempt to evaluate and regulate twenty-first-century problems. For example, we continue to assume that risk can be regulated solely through some calculus using risk analysis. This assumes that risk is manageable, that the worst imaginable incident will not occur. Moreover, current regulatory policies were designed in a different time. In the United States, the regulatory state was built up from the late 1800s and continued thorough the 1970s. This began with the regulation of transportation and the creation of the Interstate Commerce Commission and continued on through the creation of such entities as the Environmental Protection Agency. These agencies were created before computers were ubiquitous, before we had mapped the genome of numerous forms of life, before terrorism was widespread, and before the development of many of the technologies we take for granted. Not surprisingly, many of the threats we face today, including those from electronic voting, are not covered—or covered inadequately—under existing regulatory regimes. They were also created in a different time in regards to the pluralist dynamic; many groups that used to be powerful in this context, such as labor unions, no longer wield the same influence.

In fact, regulatory agencies are often in the position of both supporting and regulating an industry—think about the Department of Agriculture or the Food and Drug Administration as examples—and this creates conflict within the organization. Beck refers to this as creating the problems of "organized irresponsibility," where "institutions are forced to recognize catastrophic risks whilst simultaneously refuting and deflecting public concerns" (Mythen 2004, 64). They are also often not empowered to engage in the regulation of new issues. Instead, regulation often comes after the fact, if at all. To appreciate this in a practical manner, consider the new regulatory for chemicals scheme that is being considered by the European Union (EU). The EU's Registration, Evaluation and Authorization of Chemicals protocol requires chemical companies to register scientific data for 30,000 compounds. A subsample of 1,500 compounds that have been shown to potentially cause problems in humans will be subject to very strict scrutiny and undergo rigorous cost-benefit analysis to determine if they should be banned or left on the market. The general principle in this debate is that chemicals will have to be proved safe and will not be assumed to be safe (Cone 2005).

There is also the question of whether conventional risk assessment methods are as useful in analyzing potentially catastrophic problems as compared to studying typical risks. For example, in the case of nuclear power, Beck (1992, 29) notes that some scientists think that "risks which are not yet technically manageable do not exist—at least not in scientific calculation or jurisdictional judgment." More generally, we often know the risk of something in a specific context or specific time frame, but not in the long term. For example, science suggests that genetically modified (GM) foods are safe, but there are no data on the long-term impact of GM foods on humans, animals, or the environment. Likewise, drugs are often tested for use at a specific dosage to be given over a specific time period. Problems often arise when these drugs are then prescribed at dosages and for time periods that exceed those for which they were tested. At the same time, Sunstein (2005) notes that obsessing about unknown or difficult-to-measure risks of new technologies can lead to the development of blind spots about the risks associated with the status quo. For example, worries about GM food often ignore the risks associated with food-borne illness or starvation in countries where GM seeds would produce larger harvests than conventional crops. In short, the status quo may have risks that are also potentially catastrophic or unknown.

The failure to develop effective regulatory frameworks and to regulate risk properly can have dramatic affects on public attitudes toward government. Mythen (2004, 64–65) describes the affect of failures of risk management by expert bodies—for example, in the case of BSE in the United Kingdom—on public trust in scientists and expert institutions. Although there are subtleties to these data, such as recognizing that there can be conflicting views in science, one ill-timed event can dramatically affect public trust in key institutions like elections.

Solving the Black-Box Problem

The critics of electronic voting have come to the debate with a ready-made policy solution: the voter-verified paper audit trail (VVPAT). Largely, they have argued that voters should not have to vote on DREs at all; voters should vote on paper ballots that are counted using optical scan technology. If voters have to vote on an e-voting system, however, the critics argue that they should be presented with a printed copy of their votes as well, so that they can compare the printed ballot with the vote summary screen on the electronic voting machine. Voters would then place this printed copy of their votes into the ballot box, or it would be mechanically be placed in a sealed ballot box for the voters after verification. Should any question arise about the election outcome, these paper ballots would serve as the official vote totals, because each paper ballot was voter-verified. Also, if any part of the electronic system should crash, the paper system creates a redundant audit trail.

Not all critics of electronic voting think that the VVPAT is a meaningful solution. As we have noted, paper ballots in ballot boxes also suffer from a black-box problem: voters cannot know that the ballot they dropped into the ballot box is counted. With both e-voting systems and paper ballots in a ballot box having the black-box problem, several computer scientists and corporations have promoted cryptographic means by which voters and election officials can verify that the ballot the voter cast was cast accurately (not tampered with by the e-voting machine) and was included in the final tabulation.

A cryptographic solution would provide voters with a code that is created by combining their vote choices together using standard security mechanisms. The election official can then post the codes for all ballots that were decrypted and counted (the ballots will not record in the DRE or decrypt properly if they have been tampered with) and voters can go online or make a telephone call to determine if their vote was counted. With this technology, voters can actually verify that their vote—as cast—was counted officially. Under other paper-based voter-verification systems voters can know only that there is a paper audit trail or that they placed a paper ballot into a vote counting machine—in the case of optical scan voting. There is no way for voters to know that the vote was counted or counted accurately.

Furthermore, the criticism of electronic voting has led some states to develop new procedures in an attempt to better assess the technology's risks and to ensure that the electronic voting systems perform as desired. For example, on November 21, 2003, California secretary of state Kevin Shelley issued a series of directives related to the deployment of DREs in

California. These directives contained a series of security protocols that all jurisdictions using DREs were required to follow and came largely from the recommendations of the California Ad Hoc Touch Screen Task Force. The recommendations included:

- Parallel monitoring is required, whereby a random selection of machines of each model of DRE system is out of service on election day and then tested using a script to simulate a true election. This allows any software glitches to be identified.
- To strengthen the state's testing requirements, financial statements will be required from vendors seeking certification. Additionally, the vendor must provide the state all of the materials submitted to the federal Independent Testing Authorities (ITAs) and all ITA reports, as well as all source code and a "threat analysis."
- DRE software will be randomly certified to determine whether the code certified for use in the state is identical to the code in the machines in the state.
- Internal manufacturer security rules will require vendors to conduct background checks of software personnel before they can work on election system software. It will also prohibit voting system manufacturers from altering object code without retesting and recertification, requiring them to document a clear chain of custody for the handling of software.
- Local logic and accuracy testing rules would require that only local elections officials, not a voting system vendor representative, could conduct preelection logic and accuracy tests of a voting system.

State efforts like these can help alleviate concerns about the security risks of electronic voting devices. In the near future other states may tighten their acquisition, testing, certification, and re testing procedures for voting technologies, although these tightened security procedures need to be imposed on all voting technologies, not just on electronic voting devices.

There are a range of proposed solutions, some involving better procedures and testing, others using off-the-shelf technology (like voter-verifiable paper receipts), and others that are only in nascent states of development. Criticisms of electronic voting have to be taken seriously in order to ensure that the public has confidence in our democratic processes and is able to assess the risks of new technologies as well as to older, established technologies where the risks are more familiar. To avoid a highly bifurcated legal and regulatory framework, electronic voting and paper-based voting systems should be subject to the same stringent regulations because both can suffer from security, usability, and black-box problems.

Chapter 4

THE FRAME GAME

In the aftermath of the 2000 election, there were many calls for electronic voting. After all, many of the problems producing the 2000 election debacle arose due to the use of paper-based voting systems. There were punch card ballots that may or may not have contained a vote, depending on your view of how pregnant a "chad" could be before it gave birth to an actual vote. There were optical scan ballots where a stray mark next to a second candidate's name created an overvote, or a voter putting an "X" in the oval next to a candidate's name instead of filling the oval completely created an undervote. Watching the news in the days following the 2000 election, viewers could be forgiven for thinking that Florida's county canvassing boards were being asked to be mind readers, divining voter intent from the well-handled ballots. With the spectacle of election workers' holding the ballots up to the light to see through the chad holes, or staring and trying to determine if the line on the ballot was a vote or a stray mark, it often seemed like an Ouija board would be an equally appropriate tactic for accomplishing this goal.

These problems with paper-based voting systems in the 2000 presidential election led to many studies and reports, predominantly focused on "residuals," over- and undervotes, and spoiled ballots. There has developed a healthy academic cottage industry of studies on these questions, especially in the political science field (for a summary of the studies to date, see Alvarez, Ansolabehere, and Stewart 2005). This literature has established, especially with the additional data presented by the 2004 presidential election (Stewart 2005; 2006), a number of conclusions, some commonly accepted, others more controversial.

Among the more commonly accepted results is that punch card voting systems, especially the prescored punch card varieties (going by various trade names, including Votomatic or Pollstar), typically have some of the highest residual vote rates of any voting system used in recent elections in the United States. Although some studies found that electronic voting machines had high residual vote rates before 2000, studies using more recent data have tended to show dramatically lower residual vote rates. Scholars argue about the reasons why the more recent data indicate lower residual vote rates for electronic voting systems, debating whether it is due to new technology, heightened awareness of potential errors by voters and election officials, or better training of poll workers (Stewart 2005).

These same studies, furthermore, show that optical scan voting systems have residual vote rates that are comparable to electronic voting systems, although there is thought to be considerable heterogeneity in the residual vote estimates for optical scanning systems, because those allowing for precinct-based ballot checking appear to have much lower residual vote rates than systems that rely only on central scanning of ballots.[1]

But concerns also have arisen about whether certain classes of voters— the elderly, the disabled, those less familiar with voting technologies, language minority voters, and nonwhite or ethnic minority voters—have problems using some of these paper-based voting systems, which cause error rates to be higher for these voters than for others. If so, this presents an important normative concern, as well as potential legal problems, for decisions about paper versus electronic voting systems (Hasen 2002; Knack and Kropf 2003; Sinclair and Alvarez 2004). On this particular question, the research literature to date has been much more equivocal, with the debate largely focused on the interaction of race or ethnicity with voting system use (see Ansolabehere 2002; Knack and Kropf 2003; Tomz and van Houweling 2003; Alvarez, Sinclair, and Wilson 2004; Sinclair and Alvarez 2004; Buchler, Jarvis, and McNulty 2004). But these concerns have fueled voting-rights litigation in a series of states and have placed great pressure on election officials in other states to phase out some paper-based voting systems, especially prescored punch cards (Alvarez and Hall 2005a).

Despite this research, litigation, and legislative action to discontinue paper-based voting technologies that appear to disenfranchise potential voters, concerns about the efficacy of paper ballots have been replaced by fears of electronic voting fraud, system tampering, and system failures. The electronic voting debate has shifted from a frame of enfranchisement to a frame of fraud, errors, and potential problems of vote tampering. This reframing of the electronic voting debate fits into a historic conceit in the study of elections concerning fraud, although one ironically associated with the advent of paper ballots. The framing was also assisted, inadvertently, by a series of events (or mishaps?) associated with one company—Diebold. In a short nine-month period, the following events unfolded:

- The Diebold touch screen system was used in Georgia during the 2002 midterm elections, where several Democratic incumbents lost. This led many critics to argue that this is an example of how an election can be "stolen" by DREs.
- As was discussed previously, a version of the Diebold source code was downloaded from the Internet and examined by computer scientists. They found a series of security and concerns in the software.

- The president of Diebold stated that "I am committed to helping Ohio deliver its electoral votes to the president," a quotation that was construed as meaning he would order his computer technicians to program the Diebold machines to steal votes from Democrats.

This last item also fit into the media's ongoing view of the nation as being strongly polarized and evenly divided among "red" and "blue" states. Ever since the 2000 election, when the networks used the color red to signify the Republicans and the color blue to signify the Democrats, the media have used this distinction to describe what they view as being a highly polarized electorate. Although there is evidence that the elites—in the media and politics—are highly polarized, average Americans may not be (Fiorina 2005).

It is in this environment of polarized elites that the debate over electronic voting has occurred. An analysis of media coverage of electronic voting illustrates how the critics may have instigated a dramatic shift in the way in which the media cover this issue, with the media acting to partially stigmatize electronic voting. There has also been a strong amplification through the media of the possible risks associated with electronic voting, with little consideration given to the other side of the equation. This shift has resulted in the media adopting a consistent perspective that electronic voting is problematic.

A Tale of Two Stories

What difference does a couple of years make in the debate over voting technology? Consider the following two excerpts from the media, the first after the 2000 election, the second in 2004, just before the 2004 election. The first article is an opinion piece from the *Tampa Tribune* on May 15, 2001.

> The county commission's support for touch-screen voting was very refreshing because four members were not immediately put off by cost. Purchasing 1,400 touch-screen devices for the county's 132 precincts will cost more than $4 million, perhaps as much as $5.7 million, compared to $1.3 million for optical scanning equipment. But the cost taxpayers would avoid in the future with touch-screen voting, among other reasons, makes that system a very wise investment.
>
> Put simply, touch-screen voting is much more reliable than optical scanning, and taxpayers would not have to pay an estimated $250,000 in ballot printing costs per election cycle. Further, the touch-screen system is much simpler for voters than optical scanning—it is similar to using an ATM—and prohibits casting two ballots for one race.

The optical scanning method is similar to taking a test in school, but mistakes surely will be made. The ballots could be very lengthy, leading to voter confusion. Voters may not properly fill in an oval next to a candidate's name, or some may write the name of their choice, resulting in votes possibly being rejected.

Touch-screen voting is a paperless election method, so simple and advanced it could even encourage many more people to go to the polls—goals county officials should want. [The] commissioners are correct to turn to this technology, which, though expensive, will save money in the long run and reduce the chances of voters' clear choices being discounted.

Now compare this to comments by Paul Krugman in the *New York Times* on August 17, 2004, in one of several articles he wrote in 2004 regarding electronic voting.

Everyone knows it, but not many politicians or mainstream journalists are willing to talk about it, for fear of sounding conspiracy-minded: there is a substantial chance that the result of the 2004 presidential election will be suspect.

When I say that the result will be suspect, I don't mean that the election will, in fact, have been stolen. (We may never know.) I mean that there will be sufficient uncertainty about the honesty of the vote count that much of the world and many Americans will have serious doubts.

How might the election result be suspect? Well, to take only one of several possibilities, suppose that Florida—where recent polls give John Kerry the lead—once again swings the election to George Bush.

Much of Florida's vote will be counted by electronic voting machines with no paper trails. Independent computer scientists who have examined some of these machines' programming code are appalled at the security flaws. So there will be reasonable doubts about whether Florida's votes were properly counted, and no paper ballots to recount. The public will have to take the result on faith.

These two accounts illustrate how the framing of the debate over electronic voting changed dramatically between 2000 and 2004. As we see in the next section, such shifts in framing and tone can dramatically affect the way in which policies and policy choices are understood.

ISSUE FRAMING, POLICY DEBATES, AND THE MEDIA

The evolution in views about electronic voting over this short time frame illustrates the importance of issue framing in the debate over electronic voting. Framing is important because the context in which a story is told affects how people view the story and the related policy. In the political realm, framing allows various interests to shape the way in which a given policy is viewed. As Frank Baumgartner and Bryan Jones (1993, 5) have noted in their seminal work on policy agendas and politics, "every public policy problem is usually understood, even by the politically sophisticated, in simplified and symbolic terms. Because a single policy or program may have many implications, or may affect different groups of people in different ways, different people can hold different images of the same policies. . . . Often, proponents of a policy focus on one set of images, while opponents refer more often to another set of images." Baumgartner and Jones note further that the goal of interests on either side of a policy debate is for their image to become the dominant one, dominant to the point that they can create a policy monopoly or monopoly image. When a policy monopoly has been developed, a single image of the given policy generally exists and is accepted. Only when the policy can be reframed and a new image developed can this monopoly be broken. Breaking a policy monopoly often involves finding a new venue for making the argument; in Congress, for instance, an issue will be reformulated to come under the jurisdiction of a different committee that advocates hope will be more sympathetic to the new image.

Social scientists have long understood that the politics of images and stories are critical to the success or failure of any policy debate, and there are numerous case studies illustrating this point. William Riker (1986) examined a case regarding the location where nerve gas was to be disposed in the 1970s. The original framing of the argument by the Nixon administration was that the gas needed to be disposed of, and the selected state was the optimal choice. This framing was best for the administration, because it created a dynamic where all senators but two would support the chosen site, lest the administration open the debate back up and choose another state. The senators from the selected state, however, argued that the issue was not the location of the nerve gas disposal site. Instead, because the gas was coming from an overseas base located in a nation where the United States had specific treaty obligations, the issue was the advise-and-consent provisions of the Constitution relating to treaties. This changed the debate from one about the location of a nerve gas plant to one determining the power and prerogative of the executive vis-à-vis the Senate. Not surprisingly, this reframing was successful and the legislation was defeated. As Deborah Stone (1988, 1989) has argued in her work on policy agendas, the way in which a problem is defined is determined by competing groups offering different images and stories to explain the phenomenon in question.

One reason why images and stories are so important in policy debates is that they simplify the debate and allow it to be communicated to the broader populace through the media. In many ways, the media become arbiters in policy debates by determining which image frame they will stress among the competing frames offered by interests on the various sides of a debate. The media's power is in part determined by what they chose to cover. The media engage in a selection process, determining which issues are important relative to other issues. Some issues get covered in the media while other issues are ignored. Likewise, even among issues that get covered, the media differentiate between them, with some issues given the lead and others buried in the middle of the paper or the news show. The media thus can "prime" the public to be sensitive to certain issues and think them important.

The importance of frames and priming has been studied in many experiments (e.g., Iyengar and Kinder 1989). For example, a study of the women's movement from the 1950s to the 1990s examined the impact of four frames—feminist, political rights, economic rights, and antifeminist—on attitudes toward gender equity (Terkildsen and schnell 1997). This study found that when the media framed issues of gender equity in the context of political rights, it had a positive affect on attitudes toward gender equality. Conversely, when the issue was framed in the other ways—especially the antifeminist and economic rights frames—there was a strong negative impact on a respondent's attitudes toward gender roles, women's rights, nontraditional gender roles, and the likelihood that they would identify women's issues as being an important issue in the United States. Not surprisingly, the way in which the media discuss an issue affects the way the public views the issue.[2]

Baumgartner and Jones (1993, 60–70) identify two dimensions to the role of the media in politics: the *attention*, the volume of coverage the issue receives over a given time period; and the *tone* of the coverage, which relates to whether the frame used by the media is positive or negative. They illustrate the importance of attention and tone using the case of nuclear power. Given that nuclear power is a complex technological issue with many facets—for example, it produces clean energy and a radioactive by-product—there are many ways in which this issue can be framed, and framing and stories have been very important to the public's understanding of nuclear power. Baumgartner and Jones review the broad literature on nuclear power and note how writing on the issue has changed from a strongly positive to a strongly negative image. These studies find that the image of nuclear power has shifted largely because of changes attributable in part to how the issue has been reported in the media.[3] The lack of public understanding of nuclear power is documented in surveys that identify public misconceptions about the issue.

The nuclear power issue has seen a dramatic change in both the attention and tone dimensions in the media. From 1900 to the mid-1940s, the number of articles listed annually in the *Readers' Guide to Periodical Literature* related to nuclear power never exceeded fifty and rarely exceeded twenty-five articles annually.[4] Media attention began to increase dramatically in the later 1940s and again in the early to mid-1960s; but in the early 1960s the number of articles with a positive tone slowly drifted downward. Then, in the late 1960s the number of articles with a positive tone published each year dropped precipitously from more than 50 percent to less than 10 percent. From this point and extending through the 1970s and mid-1980s, when the analysis concluded, the number of articles with a positive tone never exceeded 20 percent and rarely exceeded 10 percent. For most of the 1970s, the number of articles with a positive tone was below 5 percent.

THE MEDIA, RISK, AND SOCIAL AMPLIFICATION

The media's critical role in the framing of issues, including the determination of which issues will receive attention in the first place and the tone of that attention, has been examined specifically in the context of understanding risk. As we noted earlier in this book, risk has become a critical aspect of the modern experience; through the media we are constantly bombarded with information and allegations of various risks, but actually analyzing risk can be difficult. As Beck has noted, even small risks can have dramatic consequences. Two scholars who have created a critical framework in the study of risk in modern society and how risks are interpreted are Roger and Jeanne Kasperson.[5] Their framework examines the social amplification and attenuation of risk. They note that risk has multiple components, including a personal component—how the risk is perceived to affect the individual—and an institutional and social aspect that affects how risks are interpreted and addressed. As they note (1996, 96), "Risk analysis, then, requires an approach that is capable of illuminating risk in its full complexity, is sensitive to the social settings in which risk occurs, and also recognizes that social interactions may either amplify or attenuate the signals to society about the risk."

One critical problem in risk analysis on the individual level is that individuals do not experience many risks directly. For example, although there are risks associated with radiation, few of us live next to a nuclear reactor or give strong consideration to the dangers associated with sun exposure because of a direct experience with melanoma. Instead, risk is understood and "experienced" through the media, and many factors affect that experience. The Kaspersons (1996) discuss the media factors noted previously—framing, attention, and tone—and note that the actual amount of information provided, along with the symbolism used to characterize the risk, helps to determine how a given risk is perceived. The medium through which information flows can also affect our risk perception, as there are differences between risk discussions that occur in mass media—television, radio, and print newspapers—compared to the Internet or specialized professional publications. Finally, the actual likelihood of a risk being serious can be inversely related to the amount of media coverage the risk receives. For example, the risk of dying from radiation exposure via the sun is much higher than the risk of dying from radiation exposure from an accident at a nuclear power plant, but the latter has received more attention, historically, than the former. Similarly, the coverage of West Nile Virus and avian flu has been inversely related to the number of deaths they cause.

To understand the political ramifications of risk amplification and the framing therein, it is beneficial to consider how the Kaspersons (1996, 98–99) describe this problem.

> Risk issues are also important elements in the agenda of various social and political groups. . . . To the extent that risk becomes a central issue in a political campaign or a source of contention between social groups, it will be vigorously brought to greater public attention, often imbued with value-based interpretations. Polarization of views and escalation of rhetoric by partisans typically occur, and new recruits are drawn into the conflict. These social alignments about risk disputes often outlive a single controversy and become anchors for subsequent episodes. Indeed, they frequently remain steadfast even in the face of conflicting information.

In short, interest groups are politically motivated to have an issue defined in a certain way. They are not likely to change their position or coalition groupings even in the face of facts or data that conflict with their views. The goal of each side in the debate is also to create an enduring perception or image of the issue that will outlast any single controversy in the debate. If they are successful in this effort, subsequent debates on this or related issues will bring to the public's mind these enduring perceptions or images. The debate over nuclear power noted previously is but one example of how such images and perceptions can remain in place for very long periods and over multiple controversies within a single policy area.

THE POLITICS OF THE ELECTRONIC VOTING DEBATE

The media's discussion of an issue, including the debate over electronic voting, is often shaped by how other (and related) issues are framed. In the case of electronic voting, its frame was shaped in part by the partisan and political context in which the issue of election reform more broadly came to the fore—a close election that had a highly partisan and bitter conclusion. In the aftermath of the Florida election debacle, many partisans on the Democratic or liberal side of the debate argued that if all of the ballots had been counted, the outcome would have been different. The analysis conducted by the National Opinion Research Center (NORC) found that, under various scenarios, the vote count would have been different had different standards been used to conduct the recount (Wolter et al. 2003). We review the NORC study in detail in chapter 6 and illustrate the myriad of issues associated with the counting of paper ballots. On the Republican or conservative side, the machinations of the local election boards and the endless counting were seen as efforts by Democrats to keep counting until they came to a result that had the Democrat candidate—Al Gore— winning. Thus, in the immediate wake of the 2000 presidential election, the initial debate about how voting technologies and election procedures had shaped the outcome of that election was politicized and partisan. This partisan theme has remained in place largely unchanged since that time.

The aftermath of the 2000 election also marks the start of the "red state–blue state" frame, and this partisan frame for national politics after 2000 may have shaped how the media covered the voting technology story and how voters perceived that story. The origins of this concept are comical; in the 1990s the television networks agreed to use the same colors for Democrats and Republicans on their election projection maps on election night. But by 2000, the networks switched colors—so if you live in a red state today you were a blue state person not too long ago!—and on election night the broad swath of the nation's south and interior were red (Republican), with the coasts and upper Midwest colored blue (Democrat). This red state–blue state division was quickly adopted by the media, especially as the 2004 election progressed. In a search in the "major newspapers" section of database LexisNexis, from 1996 to 2000, the terms red state or blue state were never used more than 6 times except in 2000, when red state was used 14 times. From 2001 to 2003, each term was used between 5 and 28 times per year, but in 2004, red state was used in 353 articles and blue state was used in 336 articles.[6]

But this dichotomy of dividing America between red and blue states has been challenged by Morris Fiorina and colleagues in the book *Culture War?* (2005). In their book, they argue that the 2000 election outcome was interpreted as meaning that the United States really comprises two nations: a conservative Christian Republican red America and a liberal, unreligious Democratic blue America. This dichotomy is intended to express a bifurcated America—potentially one in a culture war—where there is a great difference between the red and blue Americas. However, Fiorina et al. use data to suggest that individuals in red and blue states are not that different from each other across many factors, including their attitudes toward many key political issues. What is different is that there are more Bush voters and individuals who self-identify as Republicans in red states and fewer

Bush voters and more self-identified Democrats in blue states. Equally as interesting, Fiorina et al. also show that while Americans in both red and blue states generally view themselves as centrists or moderates, they see both political parties as being highly polarized. Moreover, activists on both sides have an incentive to promote polarization; after all, it is by pushing their views—and identifying the threat that exists on the other side by *those other activists!*—that they draw members and contributions to their cause.

But while the academic literature continues to debate whether the American electorate is politically polarized (e.g., Ansolabehere, Rodden, and Snyder 2006), it seems that the post-2000 discussion in the United States about voting technology is polarized, perhaps by being framed by the "red state–blue state" theme, though the nature of the polarization appears to have changed considerable since 2000.[7] What is especially interesting about the current status of the electronic voting debate is that in late 2000 and early 2001, it was paper ballots—not electronic voting technologies—that were seen as being problematic by Democrats and liberals. After all, the 2000 election crisis was prompted by an inability to count or discern voter intent on paper ballots; no electronic voting machines were used in Florida. Punch card voting and the so-called butterfly ballot were especially fingered as culprits in the debate over election reform. However, it should not be forgotten that the first lawsuit filed after the 2000 election was not filed in Florida; it was filed in Georgia, another state that used every voting technology except electronic voting. There, the ACLU and filed suit against the secretary of state stating that

> voters in some Georgia counties were ten times more likely than others to lose their right to vote because of a "fatally flawed" system that disproportionately affects people of color. . . . In its legal complaint, the ACLU called the state's voting system "a hodgepodge consisting of antiquated devices, confusing mechanisms, and equipment having significant error rates even when properly used." The state's failure to accurately record votes deprives its citizens of equal protection and due process as guaranteed under state law and by the United States Constitution, and violates the federal Voting Rights Act.[8]

This lawsuit explicitly cited voting technologies, including optical scan voting, as problematic, a finding based in part on the secretary of state's report *The 2000 Election: A Wake-Up Call for Reform and Change* (Cox 2001), that analyzed county-level data on voting. This analysis found that the residual vote rate for optical scan voting was roughly equivalent to punch cards in Georgia, with precinct count optical scan performing only marginally better than punch cards. Moreover, a precinct-level analysis that compared predominately minority population precincts in a county with predominately white precincts in the same county found relatively large racial gaps in residual vote rates, when voters used the same technology. As Secretary of State Cox told the U.S. Senate Committee on Commerce, Science and Transportation on March 7, 2001,

> We began to study undervote (in the presidential race) percentages in precincts that had black registration percentages of 80 percent or more, and compared those to predominately white precincts in the same county. We found that, across the board, undervotes are higher in predominately black precincts than in predominately white precincts in the same county. . . . But what is of greatest interest, and we think most significant as we consider equipment options, is that this undervote gap was higher . . . in counties that utilized opti-scan systems than in counties that use the punch card. . . . In this study we looked at 92 precincts with voter registration that is 80 percent or more African-American. And we compared those predominately black precincts to an equal number of predominately white precincts in the same counties. In punch card counties, the undervote in white precincts averaged 4.4 percent, while the undervote in black precincts averaged 8.1 percent, for a difference of 3.7 percent—what we are calling the "undervote gap." In counties that employ opti-scan, the undervote in predominately white precincts averaged 2.2 percent, while the undervote in predominately black precincts averaged 7.6 percent, for an undervote gap of 5.4 percent. I should point out that this higher undervote gap for opti-scan exists whether we look at counties individually or in aggregate. However we slice the numbers, in opti-scan counties, there is a greater gap in undervoting by blacks as compared to whites than there is in counties that use the punch card.

In short, all paper-based voting technologies were under attack somewhere after the 2000 election by groups such as the American Civil Liberties Union, the National Association for the Advancement of Colored People, and advocates for individuals with disabilities. But, as we will see, the politics of the voting technology debate changes considerably by 2006.

The Debate over Security and Voter Verification

Concerns about the threats associated with the use of computers in elections have been around since at least 1968. That year, Los Angeles County switched to having voters cast their choices on punch cards, which were then tabulated by computers. The *Los Angeles Times* published an article that year that expressed extensive concerns of computer scientists regarding the use of computers in vote tabulation. These computer experts noted that computer tabulation systems—which today would include almost all voting systems, from punch cards to optical scanned ballots—are vulnerable to attack, that such attacks could not be detected, that an attack would undermine confidence in voting systems and election outcomes, and that these systems should not be widely deployed. Note that, had the precautionary principle ruled at this time, we would not have had the problems with punch cards or optical scan voting in 2000. Instead, the ballots would be hand-counted, a process that would likely take weeks to accomplish given the length of ballots in many jurisdictions and the need to audit hand-counted ballots. Disabled voters would be disenfranchised, and the accuracy of the counts might be in question, but those would be the side costs of the precautionary principle. The careful student of the current debate over electronic voting security and vulnerabilities will note that such concerns today are merely the progeny of much older concerns about voting systems that are still in use today (Saltman 1988).

In the current debate over electronic voting, criticisms about the system have been promulgated from three main sources. First, there was the work of the Johns Hopkins and Rice University computer scientists who analyzed the Diebold source code for problems. Their analysis of the Diebold source code applied research principles regarding the study of computer security and writing of source code in order to identify flaws and vulnerabilities to the system. Although this work has been criticized from many sources, it did define the problems with one voting system clearly and provided a foundation for security evaluations for other systems. It also started a thorough security review process in Maryland, which uses the Diebold AccuVote TS, including the retention of security firms to evaluate the system as deployed, and to ensure that management practice in the implementation of the system mitigated the vulnerabilities that were identified by the Johns Hopkins/Rice team.

Although the Johns Hopkins/Rice research changed the dynamic of the debate over electronic voting, this research was not the sole basis for the concerns about electronic voting in 2003 and 2004. Instead, it was political advocacy by two primary organizations—Blackboxvoting.org and verifiedvoting.org—that engaged in strong media advocacy in this debate. These groups provided much of the advocacy efforts, online and in the media, that may have led to a reframing of the debate about electronic voting from a politically neutral concern about security into concerns about a possibly politically biased, completely insecure voting platform. These groups also advocated for a specific policy solution, the voter-verified paper audit trail, which would allow voters to believe that the votes they cast was saved in a paper form for auditing. These groups issued numerous press releases, held media events, and campaigned to get reporters to cover the electronic voting debate from their perspective.

MEDIA COVERAGE OF THE CONTROVERSY

In analyzing the media's coverage of the debate over electronic voting, we examined all articles in LexisNexis between 2000 and 2004, using the search terms "electronic voting" and "touch screen voting." Each article was read, and certain attributes of the article were coded, including its length, source, and whether it was a news article or an opinion piece. Then, the subject matter of the articles was examined, with consideration given to the following possible characteristics of electronic voting systems: system procurement, security, accuracy, fraud, enfranchisement, system and machine failures, counting quality, election reform, voter verification, and auditing. Finally, we examined the tone of the articles—whether the articles focused more on the benefits or failures of the system—as well as on the number of experts quoted and their affiliations. By examining these factors, we can determine how the media are defining the electronic voting debate and how this definition and framing are changing over time. Moreover, because we know when certain events occurred in the electronic voting debate, such as the release of reports regarding the efficacy of electronic voting and the creation of certain organizations involved in this debate, we can examine how these events may have affected the media's perception of electronic voting and whether the media used this new frame to discuss the risks and benefits of electronic voting differently.

We examine the media's coverage of electronic voting on three dimensions: the total number of articles, the number of opinion articles, and the number of news articles. We start with media attention to the debate, as measured by the total number of articles about electronic voting. As Figure 4.1 shows, the media's attention of the story, as measured in the volume of media coverage of electronic voting annually, follows the public's perceived interest in the story, as it relates to the national election cycles. Given the closeness of the 2000 election and the requirements in HAVA that was pushing many jurisdictions toward electronic voting, it is not surprising that the media focused heavily on electronic voting in 2004. For the period 2000 to 2004, coverage of electronic voting was higher in 2002 compared to 2001 or 2003, and more than half of all news stories— and almost 70 percent of all opinion pieces—were published in 2004.

In Table 4.1 we examine the number of news articles and opinion pieces published annually. Here, we find that most media coverage in 2000 consisted of news articles, with 86.5 percent of coverage of electronic voting in that year being news. Beginning in 2002, and extending through 2004, approximately one-quarter of all articles comprised opinion pieces. More interesting findings are apparent when we focus on two of the major newspapers that are national trendsetters in how issues are

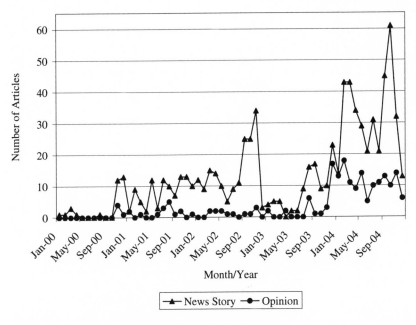

FIGURE 4.1 Media Attention to Electronic Voting

covered, the *Washington Post* and the *New York Times*. In the case of the *Post*, for the period 2000 to 2004, 62.5 percent of the paper's coverage of electronic voting that occurred during this period occurred in 2004. The *Times* coverage was even more skewed, with 76.9 percent of articles during this period appearing in 2004. The *Times* coverage was also more tilted to opinion than news; in 2004, it published four more opinion pieces (thirty-seven) on electronic voting than it did news stories (thirty-three). By contrast, 72.7 percent of the articles in the *Washington Post* in 2004 were news stories, and only 27.3 percent were opinion pieces. Nationally in 2004, 74 percent of articles about electronic voting were news stories, and 26 percent were opinion pieces.

The previous analysis illustrates how the volume of coverage changed over time and varied between news articles and opinion pieces. Next, we begin to examine the tone of the coverage, as noted in the subject matter on which the media choose to focus. In Table 4.2 we consider the subject matter that was included in each article—both news and opinion—published from 2000 to 2004. The media can frame a story about elections in different ways, and focusing on subject matter allows us to consider the evolution of topics over time. The Table starts with the issue of procurement, which was a very critical topic throughout this five-year time frame. It does, however, decrease in frequency as a percentage of all articles

TABLE 4.1
News versus Opinion on Electronic Voting Articles per Year, % (n)

	2000	2001	2002	2003	2004
News stories	86.5	86.0	69.0	73.3	74.2
	(32)	(98)	(129)	(103)	(385)
Opinion stories	13.5	14.0	31.0	23.7	25.8
	(5)	(16)	(58)	(32)	(134)
New York Times (news)	75.0	50.0	100.0	57.1	47.1
	(3)	(1)	(1)	(8)	(33)
New York Times (opinion)	25.0	50.0	0.0	42.9	52.9
	(1)	(1)	(0)	(6)	(37)
Washington Post (news)	100.0	100.0	77.8	70.6	72.7
	(1)	(6)	(7)	(12)	(40)
Washington Post (opinion)	0.0	0.0	22.2	29.4	27.3
	(0)	(0)	(2)	(5)	(15)

TABLE 4.2
Articles Defining Electronic Voting by Subject per Year, % (n)

	2000	2001	2002	2003	2004
Procurement	70.3	86.0	44.4	37.8	21.4
	(26)	(98)	(83)	(51)	(111)
Accuracy	56.8	45.6	18.7	17.0	11.2
	(21)	(52)	(35)	(23)	(58)
Enfranchisement	56.8	42.1	23.0	16.3	8.3
	(21)	(48)	(43)	(22)	(43)
Counting quality	56.8	42.1	14.4	14.8	6.4
	(21)	(48)	(27)	(20)	(33)
Fraud	10.8	14.0	2.7	48.1	40.8
	(4)	(16)	(5)	(65)	(212)
E-Voting security	5.4	17.5	11.8	62.2	69.3
	(2)	(20)	(22)	(84)	(289)
Reform	56.8	79.8	59.9	30.4	45.9
	(21)	(91)	(112)	(41)	(238)
VVPAT	2.7	19.3	2.7	43.7	52.8
	(1)	(22)	(5)	(59)	(274)
Auditability	10.8	16.7	2.1	43.0	51.4
	(4)	(19)	(4)	(58)	(267)
System failure	2.7	10.5	12.3	32.6	48.0
	(1)	(12)	(23)	(44)	(249)

throughout this period, as other issues rise to the fore. The topic that takes the place of procurement is, not surprisingly, security. Security is mentioned in 17.5 percent or less of the articles published from 2000 to 2002. By 2003 it is in 62.2 percent of all articles, and in 2004, in 69.3 percent of all articles.

This point about how security starts to dominate the debate over electronic voting creates an interesting dichotomy. In Table 4.2 we see that several positive definitions of electronic voting—as more accurate, as enfranchising voters, as a means of improving counting (especially vis-à-vis Florida and paper), and as a key to election reform—all decline precipitously after 2002 as a percent of the topics covered in articles on electronic voting. For example, in the immediate aftermath of the 2000 election, when memories about the difficulty in casting votes on paper ballots in Florida were fresh, the view that electronic voting was more accurate was a dominant theme: 56.8 percent of articles in 2000 and 45.6 percent of articles in 2001 mention accuracy. By 2002 this percentage fell to 18.7 percent and was only 11.2 percent of articles about electronic voting in 2004. Similarly, the framing of electronic voting as a means of improving counting also declined from 56.8 percent in 2000 and 42.1 percent of articles in 2001 to a mere 6.4 percent of articles in 2004. The one way in which electronic voting remains even partially positively framed through this entire period is as a means of reforming elections. In 2000 through 2002, a majority of articles in each year refer to electronic voting as a means of improving or reforming the electoral process. Even in 2004, 47.3 percent of the cases refer to electronic voting in this way, even if it may need perfecting through innovations such as improved auditability.

By contrast, the topics that dominate after 2002 are e-voting security, voter-verified paper auditing, and the possibility of electronic voting system failures. The most dramatic change is on the issue of voter-verified paper auditing. In 2000 this issue came up in a single article. In 2001 it was in 20 percent of articles, but in 2002 the topic was again only in 1.4 percent of all articles. However, in 2003, the same year as the establishment of the organization VerifiedVoting.org, the voter-verified argument is in 43.7 percent of all articles. By 2004, 75.9 percent of articles on electronic voting reference the voter-verified issue. Similarly, the argument that electronic voting might be a source of fraud was brought up in only 3.7 percent of the articles in 2000 and in 11.5 percent in 2001. By 2003 electronic voting as a source of fraud was mentioned in 48 percent of all articles and in 40.8 percent of the articles in 2004. System failures on electronic voting platforms also become a more prominent concern, moving from just 2.7 percent of the articles in 2000 to 48 percent in 2004.

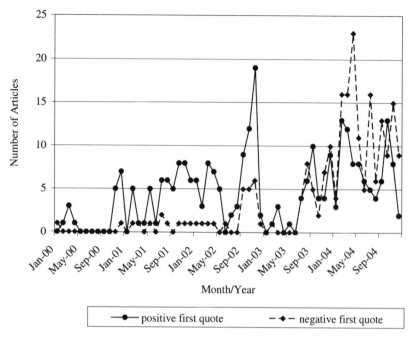

FIGURE 4.2 Tone of Articles about Electronic Voting

Given the changes in the way that electronic voting is framed from 2000 to 2004, we expect that the overall tone of the articles regarding electronic voting should also change over this same period. This is, in fact, the case. In Figure 4.2 we graph the difference between the number of electronic voting articles of positive tone and the number that were negative in tone, giving us a measure of the relative balance of tone in the stories. We see a strong positive tone in the articles regarding electronic voting from 2000 to 2002, with 86.5 percent of the articles positive in 2000 and 68.8 percent of the articles positive in tone in 2002. In 2003 and 2004, the tone is just the opposite; almost 62 percent of the articles in both years are negative in tone. As the subject matter of the articles moved toward topics that were identified with critics of electronic voting, the tone of the coverage became more negative.

Similar findings can be seen in examining the quotations used in the articles on electronic voting during this period. We can examine who gets quoted first, a person who expresses positive views on electronic voting or a person who expresses negative views, as well as the overall number of quotes from each viewpoint. This first quote is important for many reasons, including the fact that readers do not always read the entire article and the first quote often provides outside support for the

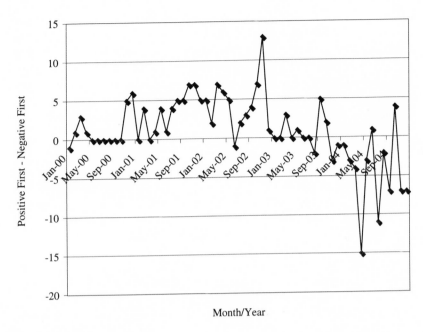

FIGURE 4.3 Tone of First Quote

frame the writer has chosen for the entire article. Again, we compute a balance measure, which provides the difference between the number of stories with a first-quote lead that is positive and the number of stories with a negative first-quote tone. This second balance measure is presented graphically in Figure 4.3.

As Figure 4.3 shows, from 2000 through 2003, the first quote is positive a majority of the time; in 2000 through 2002, more than 75 percent of the first quotes were positive. In 2004 two-thirds of the first quotes are negative, which is in keeping with the earlier data on tone and framing of the electronic voting debate. This first quote is also important in the context of the debate over electronic voting because, on average, in 2000, 2001, and 2003, there are similar numbers of positive and negative quotes regarding electronic voting in each article. This is not true in 2002, when there are 2.6 positive quotes, on average, and 1.6 negative quotes, on average, per article, and in 2004, when there is a difference between the number of positive and negative quotes, skewed toward the negative quotes. In that year, each article has 2.0 positive quotes, and 2.5 negative quotes.

Can we identify factors that might have led to the patterns of media coverage that we observe in these data? In order to see if we can account for these patterns, we analyzed media coverage over time using a simple time-series statistical model. We focus on the total number of electronic voting news stories, the total number of opinion stories, the positive or negative story tone, and the positive or negative tone of the first quote in each story, from 2000 through 2004 (the same data we have just examined above graphically). In this analysis, we considered how a variety of factors and events shaped media coverage. First, we considered how media coverage one-month prior helps to shape the current month's coverage. Second, we examined how the following salient events since 2000 might have affected media coverage when they occurred:

- The Florida 2000 election (November), which sparked the controversy over election processes and technology
- The Florida 2002 primary (September 2002), which we suspect sparked renewed interest in media coverage of voting technology issues
- The passage of the Help America Vote Act (October 2002)
- The November 2002 election
- The initiation of the VerifiedVoting.org (January 2003)
- The issuance of the first major report criticizing the Diebold voting system by Avi Rubin and his colleagues (July 2003)
- The cancellation of the SERVE project by the Department of Defense (February 2004)
- The preelection coverage of the November 2004 general election (October 2004)
- The November 2004 election

Instead of providing the statistical details, we present a simplified explanation of the results here. First, we consider the amount or volume of media coverage: this analysis indicates that past media coverage of electronic voting has a predictive effect on the current month's coverage. There are also specific events that appear to help predict media coverage of electronic voting in news stories. The most important events, listed in order of the relative impact they had, are the cancellation of the SERVE project in early 2004, the preelection coverage of the 2004 presidential election in October 2004, the Florida 2002 primary, the 2002 general election, and the November 2004 general election. With the exception of the 2004 general election, which represents the end of the media attention span for the election, all of these events served to increase media news stories about electronic voting.

We find that far fewer events predict media opinion stories about electronic voting. Only the cancellation of the SERVE project and the November 2004 general election have a positive impact on the number of opinion stories. These two events led to a higher number of media opinion stories about electronic voting, controlling for the other events that occurred and the past month's history of media opinion stories on this issue.

When consider our two measures of the tone of media coverage, we find that it is more difficult to predict the tone of media coverage with this simple time-series model than to predict the overall amount of coverage. We again find that the tone of last month's media coverage helps to predict the current media tone leading to the inference that media tone in current coverage of the electronic voting issue is influenced by the past month's tone. As for events that predict tone, we see that the overall tone of media coverage was affected by the Florida 2000 election and the November 2002 election. Both of these events had strong and positive effects on media tone. The other events did not help to shape the tone of coverage. The story is different when we look at the tone of the first quotes. Here we see that the November 2002 election, the preelection coverage in October 2004, and the passage of HAVA in October 2002 all had a significant and positive impact on the tone of the first quote. However, we also find that the November 2004 election is different. Here, we see that in the immediate days before and after the November 2004 election the first-quote tone was decidedly negative.

In conclusion, these time-series models have help us better understand some of the systematic factors behind media coverage of electronic voting in 2004—and thus to better understand the "frame game." We find that the media coverage has a strong historical component—that is, how the media covered electronic voting in the previous month had a uniform effect on subsequent media coverage of electronic voting. We also found that the November 2002 and November 2004 elections each had significant effect on media coverage of electronic voting in three of four models, implying that the proximity in time to these two general elections, where vote casting was clearly on the minds of the media and public, were the most consistent events predicting media coverage of electronic voting during this time. We see evidence that the cancellation of the SERVE project is correlated in time with the number of media news and opinion stories, though not with the tone of their coverage.

RISK, THE MEDIA, AND ELECTRONIC VOTING

The media coverage of the debate over electronic voting reflects the success of critics of electronic voting in defining the way in which the debate played out after the initial reform period that followed the 2000 election and lasted through the passage of HAVA in 2002. The critics of electronic voting were able to stigmatize, at least in part, this voting technology, especially by arguing that these systems were not appropriate without voter-verified paper audit technology. Importantly, this reframing occurred in 2003, just before the greatly increased coverage of voting technology that occurred concomitant with the presidential election in 2004 and the first presidential race since the 2000 debacle. Critics of electronic voting shifted the coverage of this debate in the year that had more coverage of electronic voting than in the four previous years combined.

If we consider the two factors that Baumgartner and Jones found to be important in issue framing, attention and tone, we see that the two joined most opportunely for critics of electronic voting. The height of attention, the 2004 election season, was also the point at which the tone of coverage had shifted from primarily positive to primarily negative. This shift can be seen in the topics that were covered, such as security and the need for voter-verified paper audit technologies, as well as in the topics that were no longer of focus, such as enfranchisement of individuals with disabilities or minority voters. Moreover, the overall tone of articles was now more negative, which can also be seen in the movement to leading with a negative quotation about electronic voting, which was a shift from earlier coverage, which often mentioned the potential problems with electronic voting but tended to lead with the positive aspects of this technology.

The concerns of sociologists of the risk society have often been that risks are downplayed until it is too late. In the case of electronic voting, the media amplified the risk concerns that were being expressed by some computer scientists, as well as by advocates such as VerifiedVoting.org. These organizations were able to shape the debate over electronic voting quite successfully. In fact, the debate was used for political purposes, with groups such as MoveOn.org using the potential for electronic voting fraud to steal the election for President Bush as a mobilizing technique. Unfortunately, the scientific arguments regarding how to evaluate and implement the best voting technologies and auditability systems were lost in the debate.

Chapter 5

ONE STEP FORWARD, TWO STEPS BACK

When we published *Point, Click, and Vote: The Future of Internet Voting* in January 2004, we had little idea that we should have been publishing the book in Europe, not the United States. In the book, we discussed the potential benefits of Internet voting and laid out a road map for how policy makers could conduct experiments to learn how Internet voting could be utilized to address the voting needs of special populations, especially military personnel, overseas civilians, individuals with disabilities, and similar groups who have had difficulty voting under the current voting process. The road map we lay out in the book is being followed, just not in the United States. Instead, it is in countries like Estonia, Switzerland, the Netherlands, and France that e-voting experiments are being conducted.

In 2002 the future of Internet voting seemed to be cautiously bright, largely because the history of Internet voting efforts had been positive. The Department of Defense had carried out a successful Internet voting trial in the 2000 general election, and the Democratic Party of Arizona had also carried out an Internet voting trial in the 2000 primary election that received positive media coverage. Likewise, successful Internet voting trials were being held throughout Europe, including the United Kingdom, France, and Switzerland. In these trials, there had not been any documented security problems, the central critique of Internet voting, and evaluation efforts indicated that participants in these trials had enjoyed the online voting experience. Although the trials had not boosted turnout as some had hoped—increased turnout was an explicit goal of the experiments in the united kingdom—the experiences were problem-free.

With the invasion of Iraq and the deployment of more than 100,000 forces in Iraq and the continued deployment of thousands of U.S. forces in Afghanistan, the 2004 election had the potential to be more significant in the history of the American electoral process than most knew because of a requirement by Congress on the Department of Defense to deploy an Internet voting system to facilitate voting by military personnel in this important election.[1] This system was to have addressed the historical problems faced by military personnel, their dependents, and overseas citizens in casting ballots using the current paper-based system. As many as 200,000 voters were initially expected to cast ballots on the system, which would have made this effort one of the largest and most extensive experiments with Internet voting conducted.

This Internet voting effort never came to fruition. Instead, the Department of Defense's Secure Electronic Registration and Voting Experiment (SERVE) was canceled. However, in 2004 the Michigan Democratic Party held its Democratic caucus using Internet voting as one voting method. As we will discuss, SERVE was a direct casualty in the general debate over electronic voting. In fact, there is some reason to believe that the fierceness of the debate over electronic voting, for some critics, is a result of their desire to ensure that Internet voting never becomes commonplace. One of the most fascinating things about the SERVE project is that this program was an effort to conduct an experiment and to learn about a voting technology and its effectiveness before it was fully deployed. This willingness to experiment is unusual in the United States and is something that differentiates it from European counties.

THE IMPORTANCE OF PILOT TESTING IN ELECTIONS

In the United States, changes to election administration—from the implementation of new voting systems to the implementation of expanded early voting—are typically conducted in large-scale, nonscientific tests. A change is made to the law, to procedures, or to technology, and then is implemented in the next election. Often these implementations are undertaken without much attention paid to whether they achieve any necessary and stated purpose. Sometimes, sweeping changes are implemented, and policy makers are interested in determining the extent to which their implementation has influenced some outcome variable. Unfortunately, the policy maker is quite constrained in the ability to learn much from the "test" because it has not been undertaken in a controlled manner.

This American model of electoral reform implementation can be contrasted with the more deliberate model of implementation that has been followed in Europe, where election reforms are more often pilot-tested under controlled conditions before being implemented in a large scale. This has been especially true in the area of electronic voting technologies and convenience voting in the United Kingdom, France, and Switzerland. The United Kingdom provides a model of how such pilot testing can be done, in contrast to the way in which such tests occur in the United States.

In the United Kingdom from 2002 to 2004, the Electoral Commission and the Office of the e-Envoy conducted a series of experiments, or pilots, examining various election reforms and new voting techniques. The pilots had several goals. As the commission wrote in its report *Modernising Elections,*

> The pilots took place against a backdrop of seemingly irreversible declining participation in local government elections and the substantial drop in turnout in June 2001 general elections. . . . However, turnout was not the only, or even primary, goal of the pilot schemes. Some were looking for administrative efficiency gains; others wanted to be involved in the state or the process of developing electronic voting mechanisms robust enough to win public credibility.[2]

Given the varied goals of the pilots for the stakeholders involved in the process, the pilot tests had a multipart evaluation component. The evaluation methodology, in practice, had two components. First, there was an analysis of the pilot scheme, as implemented in a specific locality, to determine its affect on factors such as turnout, election administration, and fraud. Second, there was an analysis of public attitudes toward the experiment in question by survey research. The evaluation criteria used in the pilot schemes are a part of a larger model of evaluating program effectiveness. This process begins with the identification of evaluation criteria—some of which are contained in the statute creating the Electoral Commission and its mandate to conduct pilots—and are supplemented by criteria from other sources. The various components of the pilot are then identified and implemented, and the pilot is evaluated. Finally, the pilot program has a very strong publication and evaluation component, which ensures that the findings of these pilots are widely disseminated and are used as the basis for decision making regarding what types of future pilots to conduct.

The U.K. pilot schemes for election reforms benefit from following a strong piloting and evaluation strategy. Not only were the pilot schemes implemented in jurisdictions across the United Kingdom in local elections, but detailed evaluation reports were produced on each pilot. Then an overall report was issued after each set of pilots was completed to provide policy makers with a plan for how to move forward with election reforms, based on the social science research evidence at hand. Finally, these reports and much of the evaluative data used in the pilot tests are made available to other researchers, policy makers, and the public so that additional analysis can be undertaken.[3]

There were three sets of pilot schemes implemented between 2002 and 2004. There were thirty individual pilots in 2002, fifty-nine pilots in 2003, and forty-seven pilots in 2004. Table 5.1 lists the types of pilots that were conducted each year and illustrates the benefits of conducting pilots. The pilot process allowed for a wide range of technologies to be tested but in carefully chosen settings. By piloting on a small scale and providing

TABLE 5.1
Pilot Types by Year

	2002	2003	2004
Pilots	30 localities	59 localities	4 regions
Postal			
All-postal	15	35	4
Postal channel	9	4	Scotland, Wales, London
Electronic voting mechanisms[a]			
E-counting	14	8	
E-voting	7	8	
Telephone voting	3	12	
Text message voting	2	4	
Internet voting	5	14	
Interactive television	0	3	
Other			
Early voting or longer hours	9	13	

Note: The 2002 Pilots are discussed in detail in the report "Modernising Elections: A Strategic Evaluation of the 2002 Electoral Pilot Schemes," http://www.electoralcommission.org.uk/files/dms/Modernising_elections_6574-6170_ E_N_S_W_.pdf; the 2003 pilots in "The Shape of Things to Come." http://www.electoralcommission.org.uk/files/dms/The_shape_of_elections_to_come_ final_10316-8346_E_N_S_W_.pdf; and the 2004 pilots in "Delivering Democracy? The Future Of Postal Voting." http://www.electoralcommission.org.uk/files/dms/DeliveringDemocracyfinalcomplete_ 16306-10935_E_N_S_W_.pdf.

[a]No e-voting pilots were conducted in 2004, only postal pilots.

multiple channels for voting in cases where new technologies were being implemented, the risks associated with a problem in any one voting system being piloted were greatly minimized. Once the small-scale pilots in 2002 proved successful, these pilots were expanded in 2003 so that more information could be collected on the critical factors of success identified at the outset.

Electronic voting pilots were held in 2002 and 2003. These pilots tested a wide range of electronic voting technologies, including voting over the Internet, on precinct-based touchscreen machines, over text messaging systems, via the telephone, and using interactive digital television services. These trials were all successful in that they produced useful data and helped to inform ongoing election reform debates. In fact, respondents in focus groups state that "voting using the internet is tacitly accepted by most as 'the way forward' (at least in conjunction with other methods). Some see it as a logical, and perhaps even inevitable development, especially in the context of the younger generation's perceived preference for communicating electronically" (Electoral Commission 2003, 66). Also, when given a choice between paper ballots and new electronic technologies, voters chose the electronic technologies. Compared with the postal option, however, the electronic channels, while beneficial for many voters, did not result in the large boost in turnout that the Electoral Commission had hoped.

The Electoral Commission was able to learn quite a bit about the issues associated the deployment of these systems—for example, issues associated with interacting with vendors, such as contract negotiating, quality assurance efforts, and project management, and the related issue of scalability and cost of scaling. These systems require professional support and relatively high usage to justify the costs, and developing a business model for elections is needed to make such systems effective. Survey data from the pilots allowed the commission to see the concerns that some voters had with the potential for fraud in the system, a finding that occurred with the postal voting as well. By piloting, and using the pilots to develop a plan for future action, the commission has been able to develop a new timetable for further testing electronic voting systems and develop a plan for further investments in these technologies.

The commission also focused on two issues that are far too often overlooked in the United States—project management capacities and public relations. Its report on postal voting noted that postal voting requires the development of new administrative capacities by election officials, the political parties, and third-party contractors in order for it to be successful.[4] A different management model used for large-scale postal voting involves outsourcing services such as printing and processing ballots for delivery to voters. It also required election officials to address the issues associated with contracting and procurement. On the issue of public relations, the commission explicitly has focused throughout the pilot process on the importance of educating voters about pilots and about changes in election processes and procedures. In the United States, too often these changes are not well communicated either to voters or to the candidates in elections.

Pilots are but one feature of the Electoral Commission's research efforts related to election reform. It also engages in a regular pattern of surveys, focus groups, and policy analyses in order to inform its decision making. For example, in just the past two years, it has commissioned studies on ballot security, electronic tabulation, and vote registers. It has also conducted survey research on issues such as political engagement, reasons for voting and nonvoting, the success of the pilot schemes, gender and political participation, and perceptions on fraud.[5] In short, its work is well informed regarding the public's perceptions as well as the realities on the ground. Interestingly, the United Kingdom has become much more concerned about ballot security and the security of all voting platforms, including electronic voting, because of allegations of fraud associated by postal voting—what Americans would refer to as by-mail absentee voting. These fraud allegations involved paper-based voting technologies being stolen or manipulated in ways that undermined public confidence in the voting process.[6]

The Electoral Commision model is not unique to the United Kingdom but is reflective of a different model of election reform in Europe. Table 5.2 provides an overview of electronic voting in Europe; it shows that other European nations are following a similar vein in piloting voting technologies. As of late 2006, a total of eight nations have conducted real remote Internet voting pilots. Two pilots that closely mirror the efforts that have been attempted in the United States are the French and Dutch pilots for allowing expatriates to vote online. However, more extensive efforts have also been attempted. In addition to the experiments with remote Internet voting in the United Kingdom, the Estonians first piloted Internet voting in local elections and then expanded their efforts to cover the entire nation in recent elections. The efforts in the United Kingdom, Netherlands, and Estonia are all a part of political and legal frameworks for promoting experimentation and innovation in the area of elections. The Dutch experiments in e-voting are even authorized under an Experiments Act.

The political framework in which e-voting is being conducted in Europe is generally one that is characterized by deliberative consideration of the issues associated with moving to e-voting. In Germany, a phased approach is planned, with Internet voting to be tested first in nongovernmental elections, then local elections, with the goal of having an online national election later. In Switzerland, Internet voting was pilot-tested in Geneva in January 2003 without incident. The Swiss also hired a team of "white-hat" hackers to try to break into their security system over a three-week period—the system was online to voters for only two days—but the hackers failed (Alvarez and Hall 2004). The Estonian government plans for remote e-voting are carefully integrated into a national program that promotes e-government services generally. The Estonia government promotes the evaluation of its voting system and routinely invites international scholars to observe and study its elections.[7] Over the next four years, remote e-voting experiments will be conducted in more national elections across Europe.

The European model of pilot testing can be contrasted to the U.S. model, where tests are rarely conducted in a comprehensive and scientific manner. One example of such a pilot in the United States is the case of Georgia's move to electronic voting. There, the state engaged in a set of trials, where each approved voting system was used in a municipal election. Voters were surveyed after using the system, and statewide surveys were conducted to gauge the public's views toward the move to electronic voting. These data, along with data from the pilots, were used to select a voting machine that was then deployed statewide.[8] Similarly, the city of Alexandria, Virginia, conducted a trial to test two voting systems before selecting one for purchase.[9] This trial consisted of testing the new voting technology in a single precinct, surveying voters at that precinct regarding their attitudes toward the new technology, and collecting other data on system performance. These data were compared to data from a "sister precinct" that had similar demographic characteristics.

The 2004 Secure Electronic Registration and Voting Experiment (SERVE), built on the Internet voting proof-of-concept trial conducted by the Federal Voting Assistance Program in 2000, had a strong evaluation and research component that would determine the effectiveness of key aspects of the pilot. Through the use of surveys and data collection from matched precincts, it would be possible to determine if SERVE improved the access to voting. Because it was a relatively small-scale experiment—targeting less than 100,000 potential voters in an election that would have over 100 million eligible voters—it would allow us to learn for the first time about the way in which this unique technology functioned within the political and administrative culture of the United States. In the end, a small but vocal segment of the scientific community opposed the use of scientific experimentation in voting systems and technologies.

SERVE: THE PROJECT THAT WASN'T

Entering the 2004 election cycle, the largest Internet voting trial for American voters was to have been conducted by the Federal Voting Assistance Program (FVAP), which administers the Uniformed and Overseas Citizens Absentee Voting Act (UOCAVA) for the U.S. Department of Defense.[10] UOCAVA voters are the more than 6 million military personnel, their dependents, and overseas citizens who are eligible to vote but who historically have had difficulties voting using the paper voting process. Section 1604 of the Fiscal Year 2002 National Defense Authorization Act (Public Law 107-107) directed the secretary of defense to carry out an electronic voting demonstration project in 2004 in concert with

TABLE 5.2
Electronic Voting in Europe

Country	Research Groups	Government Program and/or Working Groups	Real Remote E-Voting (in Pilots)
European Union			
Austria	E-Voting at (University of Economics and Business Administration Vienna)	"Arbeitsgruppe E-Voting" in the Ministry of the Interior (BMI)	National referendum in 4 municipalities (September 2004)
Belgium			
Denmark			
Estonia		E-voting project for the local level (since August 2003)	Talinn Referendum (2005)
France	Working group (forum des droits de l'Internet) issued road map for implementation of e-voting in France		Expatriates living in U.S. could vote for the "High Council (French Citizens Abr (CSFE) (March 200 E-voting in non-political elections, such as chambers of commerce
Germany			Several pilots since 1999
Great Britain		Electoral Commission	Piloting at a large sca
Ireland		Government-sponsored independent "Com
Italy			
Netherlands			Broad e-voting pilot expatriates in the elections (2004)
Portugal			
Spain			Catalonians abroad regional parliame (November 2003 municipalities e-parliamentary el (March 2004)
Switzerland			2003 and 2004

Government Level (of Pilots)	Remote E-Voting Tests	E-Voting at Polling Stations	Legal Framework	Political Framework	Motivation/Goals for Introducing E-Voting	(First) Target Groups
			European Signature Directive (1999) as basis for digital signature cards			
Federal	Voting test parallel to the president's elections (April 2004) (eligible voters: students of WU Vienna)		E-Government Act 2004: concept of Bürgerkarte (digital signature card) that can eventually be used for voting, ATM withdrawal, etc.		More information and participation of the citizens = more democracy	Non-political elections (universities, chambers of commerce unions, etc.): groups with high Internet access rates
		X				
	X	X				
Local			Local Government Council Election Act, Riigikogu Election Act, Referendum Act			
National, nonpolitical	Tests					Expatriates
Nonpolitical elections only (e.g., universities)	X	X	Governmentally commissioned set of technical and operational requirements for online voting systems (April 2004). Legal restrictions on e-voting at political elections.			Nonpolitical elections
Local						
on Electronic Voting"			Plans (since 2000) to e-vote in 2004 were finally not introduced			
		Tests				
National		X	Elections Act (excludes remote E-voting). Experiments Act to allow experiments			Expatriates
	X	X				
Regional, national	X	X				Expatriates
Local			High number of referenda			

Note: Definitions by the Council of Europe: e-election = a political election or referendum in which electronic means are used in one or more stages; e-voting = an e-election that involves the use of electronic means in at least the casting of the vote.

state and local election officials. The project was to be called the Secure Electronic Registration and Voting Experiment, or SERVE.

SERVE was intended to determine whether electronic voting technology could improve the voting participation success rate for UOCAVA voters. To achieve this goal, the project had an extensive program evaluation component. One success criteria would be the participation change of several subcategories of UOCAVA voters, such as activated Guard and Reserve units or federal employees overseas. A second objective was to assess the potential impact on state and local election administration of an automated alternative to conventional by-mail absentee registration and voting. Achieving these research objectives depended on having broad participation by voters and by state and local jurisdictions.

SERVE was initiated during a time of transition among election officials, and recruiting states to participate was difficult for many reasons. Some states needed to pass enabling legislation to participate in such a trial. Other states suffered under staffing and budgetary constraints and were concerned about additional workload. Finally, HAVA and the 2000 election caused turnover among state and local election personnel and affected their ability and interest level in participating. Table 5.3 lists the states and localities, as of January 20, 2004, whose participation in SERVE was pending. The goal was to have up to 100,000 voters participate in the SERVE project, with many if not most of them using the system in both 2004 primary and general elections.

FVAP had already run an Internet voting trial—the 2000 Voting over the Internet (VOI) pilot—which we studied in *Point, Click and Vote.* This pilot allowed eighty-three UOCAVA voters from five states to cast ballots online in the 2000 general election. SERVE's original objective was to build on the lessons learned from VOI, and to allow voters to register and vote using any computer with Internet access any time, anywhere. It would also have allowed voters to register from one location and vote from another without having to notify an election official. This flexibility of access and location independence could have been especially well suited to the circumstances of uniformed service voters.

The SERVE system was ambitious; in addition to providing a fully functional registration and voting system for distributed voting, it was also designed to provide all the functions of a complete UOCAVA-voting administration system staffed by election officials participating in the project, a system that would have been used by local election officials as an adjunct to their existing local systems. These local systems, such as voter registration, would continue to be the system of record. The SERVE system included the following capabilities for election officials: identification and authentication enrollment;[11] system setup; voter registration and absentee ballot request application processing; election definition, ballot

TABLE 5.3
States and Counties That Were Likely SERVE Participants

	South Carolina (cont.)
Arkansas	Calhoun
Benton	Colleton
Boone	Florence
Craighead	Greenville
Crawford	Lexington
Faulkner	Orangeburg
Jefferson	Pickens
Pulaski	Richland
Washington	Spartanburg
Florida	Sumter
Bay	Williamsburg
Clay	York
Miami-Dade	Lancaster
Okaloosa	Greenwood
Orange	Chester
Osceola	Cherokee
Hawaii	Marlboro
Hawaii	**Utah**
Honolulu	Davis
Kauai	Sanpete
Maui	Tooele
North Carolina	Utah
Craven	Weber
Cumberland	**Washington**
Onslow	Cowlitz
Pasquotank	Island
Wayne	Kitsap
South Carolina	Pierce
Aiken	Spokane
Anderson	Snohomish
Beaufort	Thurston

conversion, and proofing; voted ballot receipt and reconciliation; ballot
tabulation; and reporting. The following capabilities were provided for
voters: identification and authentication enrollment, voter registration and
absentee ballot request application submission, status checking, absentee
ballot availability notification, absentee ballot delivery and voting, ballot
choice confirmation, and voted ballot return.

The system architecture consisted of a central server environment that hosted all the voter functions and all the election administration functions except ballot tabulation. A dedicated laptop was to have been provided to each participating jurisdiction for the function of downloading the anonymous voted ballots, decrypting the ballot data, and tabulating the results. To use this system, authorized local election officials (LEOs) could log on to the central host from nearly any Windows-compatible computer in their office, authenticate themselves using a digital signature,[12] and access any function for which they were an authorized user. Similarly, any voter who had been issued a SERVE digital signature could access any of the voter functions from any Windows-compatible computer. No special software was required to be installed on any LEO or voter computer.

Full participation in SERVE would have been a three-part process for voters. First, interested persons would have logged onto www.SERVEUSA. gov and submitted an application to register to participate. If they met the qualifications—which required either having an appropriate military digital certificate or having paperwork reviewed by an authorized individual with an approved digital certificate—they would receive a digital signature that would allow them to authenticate themselves to the SERVE system. Individuals would then register to vote with their election jurisdiction. Each registration request would be reviewed by the LEO to ensure that potential voters in fact were eligible—meaning they were UOCAVA-covered individuals in a jurisdiction participating in SERVE. If voters were eligible to vote, they would be put into the appropriate LEO's voter registration system. When it was time to vote, voters would again use their digital certificate to log onto the system, receive the correct ballot, have built-in mechanisms to check their ballot and accompanying materials for common mistakes, "sign" the oath, and confirm their choices before submitting their ballot.

Because it was being used in a general election, SERVE created several unique accreditation and certification issues for participating states. First, the 2002 Voting System Standards (VSS) applied to the voting and vote tabulation functions, but no standards exist for voter registration and other system functions. Second, the 2002 VSS did not contain standards for a Web-enabled, Internet-based system, and there was no significant precedent for applying the VSS to this technology. Third, the Department of Defense has its own standards for systems used by the department, even if the department is not a system user. (The department had no ability to access the SERVE system once it was operational.) This put the FVAP, the system developer, and the independent review body in a position of having to develop a certification process for this system.

SPRG and SERVE Termination

We were the principal investigators of the only research evaluation contract issued by FVAP and the Department of Defense to evaluate the efficacy and security of the SERVE project. As part of this evaluation effort, in late 2002, when the project was starting, discussions between us and the FVAP project manager began about the establishment of a peer review group that could provide external perspective and input into the SERVE registration and voting system's architecture, security, usability, and technical development. These discussions led to the development of a Security Peer Review Group (SPRG), and more than a dozen computer science and information technology experts were invited to participate in the SPRG. The early conception of the SPRG was that the members of the review group would provide initial input to the system development team about technical issues and would be involved in postelection evaluation of the technical issues of the system's performance. But the SPRG was one of two groups examining the security of the SERVE system. A second group of government information assurance experts from various federal agencies—including the CIA, the National Security Agency, and the Department of Defense—worked in parallel to the SPRG, examined the SERVE system, and gave input to the system development team.

The first meeting of the SPRG group was on July 8 and 9, 2003, on the Caltech campus in Pasadena. It included some of the invited external experts, as well as representatives from the FVAP who were overseeing the project and key members of the project design and development team. The SPRG was extensively briefed at this first meeting on the SERVE project and given overviews of the system's functional requirements, technical architecture, and security design. The SPRG external reviewers were given the opportunity to engage the design and development team in question-and-answer sessions. The development team modified the security of the system after this meeting to address the wealth of comments and questions offered by the SPRG.

A second meeting of the SPRG was convened November 10–12, 2003, in Reston, Virginia, where the SERVE development team was primarily located. This meeting focused on aspects of the SERVE system that had evolved since the July SPRG meeting, with additional briefings from key FVAP and development team personnel. In addition to the briefings, SPRG members were given the opportunity to undertake an extensive "test drive" of the SERVE system. At that point in time, many of the basic modules of the SERVE registration and voting system, from the perspectives of election official and voter, were far enough along the development path to be used in a "mock election" setting by SPRG members. This "test drive" resulted in a substantial amount of discussion by SPRG members into questions about usability and system architecture. There was also a substantial discussion of system design requirements, how the project was evolving to meet those requirements, and what additional steps might be necessary to improve the system's security and usability.

Unbeknownst to the FVAP, the SERVE development team, and most of the participants in the SPRG process, four SPRG participants worked on a report criticizing the SERVE project following this second meeting in Reston, calling for project termination. This report, released to the media and public in late January 2004, led to negative media coverage of the project and ultimately led Deputy Secretary of Defense Paul Wolfowitz to sign a memorandum blocking the use of the SERVE system for the November 2004 election. As a Defense Department spokesperson summarized the impact of this memorandum (which was not released directly to the public), "The department has decided not to use the SERVE program in the November elections because of our inability to ensure the legitimacy of the votes."[13]

Interestingly, the report did not directly criticize the SERVE system or its architecture. Instead, the problem with Internet voting is with the Internet itself. To quote the authors:

> The real barrier to success is not a lack of vision, skill, resources, or dedication; it is the fact that, given the current Internet and PC security technology, and the goal of a secure all-electronic remote system, the FVAP has taken on an essentially impossible task. *There is no good way to build such a voting system without a radical change in overall architecture of the Internet and the PC,* or some unforeseen security breakthrough. (p.83; emphasis added)[14]

In short, the authors posit that any Internet transaction—from buying a book to e-filing your taxes—is dangerous when done online. The authors are also open about their biggest fear, which is not that the system will not work but that the system might have worked well. Given that there has never been a known failure yet in a public election using Internet voting, this might have been the case.

These four authors of this report, David Jefferson, Avi Rubin, Barbara Simons, and David Wagner, put a copy of their report on a Web site with some supporting materials, including a press release and a copy of a *New York Times* article discussing their report.[14] At the heart of their report is the following argument. First, they assert "Our analysis is therefore premised on the following principle: *At the very least, any new form of absentee voting should be as secure as current absentee voting systems*" (p.8; emphasis in original). Based on this principle, they then go on to assert that unlike the current methods available for UOCAVA voters to cast absentee ballots, the SERVE system was less secure than those methods, because they assert that the SERVE system could have allowed for attacks of a larger magnitude. In their words, "It must be essentially *impossible* that any such large-scale attacks go undetected; or that such an attack might be so easy and inexpensive that a single person acting alone could carry it out; or that the perpetrator(s) of such an attack might never be identified; or that such an attack might be carried out remotely, from foreign soil, possibly by a foreign agency outside the reach of U.S. law, so that the attackers face little or no risk" (p. 8; emphasis in original). The rest of their report goes on to provide a list of previous critiques of Internet voting, some of them articulated by some of these authors.[15] In short, these SPRG members were opposed to Internet voting in principle, regardless of how the system was designed or implemented.

Of course, it is now impossible to know whether the authors of this report were accurate, as the SERVE project was not allowed to proceed in the 2004 election. Our perspective is that the central argument in this critique was overly general, ignored the reality of UOCAVA voting, and ignored what would have been a broad array of project, procedural, and architectural details of the SERVE registration and voting system, which in all likelihood would have minimized or mitigated their concerns had the system been used in the planned trial.

Estimates indicate that there are between 6 million and 7 million Americans who are overseas, who are in the Armed Forces, or who are dependents of Armed Forces members. These American citizens include soldiers stationed in places like Iraq and Afghanistan, currently fighting the war against terrorism; they include missionaries working in remote regions of the world; they include younger Americans studying abroad; and they include Americans who work overseas, building economic opportunities in the global economy. These citizens face very significant hurdles when they try to participate in the American democratic process, especially those citizens who are in very remote, very dangerous, or even highly sensitive locations. This is true even given the good-faith efforts undertaken by many governmental agencies, election officials, employers, and other groups to help Americans overseas cast ballots (GAO 2001a).

As the debate over the SERVE project heated up in early 2004, the United States was in the midst of a major military mobilization in two different theaters, Afghanistan and Iraq, as well as fighting the general war on terrorism in other parts of the world (Alvarez, Hall, and Roberts, 2007; GAO 2001a). There was great concern in early 2004 that the many thousands of American military personnel, and citizens supporting their efforts overseas, might have difficulty voting in the 2004 presidential election. An April 2004 GAO report added fuel to these concerns about logistical difficulties associated with Operation Iraqi Freedom. It found that, for many military personnel, mail was never delivered or sometimes received weeks, or even months, after it was sent. By contrast, U.S. personnel in Iraq and Afghanistan have a high level of Internet connectivity and access to computers because of its importance for morale and communications to families at home.

Data from numerous studies and analyses conducted since the 2000 election show that civilians living overseas and personnel in the uniformed services have a difficult time participating in the electoral process using the current paper-based absentee voting system. An examination of absentee voting in California found that UOCAVA voters were roughly two times more likely to not return a requested absentee ballot and approximately three times more likely to have that ballot challenged, relative to non-UOCAVA voters (Alvarez, Hall, and Sinclair 2005). Likewise, the *New York Times* conducted an independent examination of late overseas absentee ballots received in the 2000 Florida election after November 7, 2000, and examined by canvassing boards between November 17 and 26.[16] There is was an extremely high error rate in how ballots were handled; many voters cast ballots that were not counted, while others that might not have qualified as acceptable ballots were counted.

In separate reports, the United States General Accounting Office and the Department of Defense Inspector General also found that the current UOCAVA voting process is very cumbersome, resulting in many voters being disenfranchised.[17] One problem is that the rules governing UOCAVA voting vary by state, which can cause confusion. Military UOCAVA voters especially suffer from serious problems when they wish to vote because of logistical difficulties in getting ballots to soldiers moving across and throughout an operational battlefield. The paper-based process is also a source of many problems. As the GAO (2001a) noted, "[M]ilitary and overseas voters do not always complete absentee voting requirements or use federal forms correctly. . . . County officials said that problems in processing absentee voting applications arise primarily because voters do not fill in the forms correctly or do not begin the voting process early enough to complete the multiple steps they must take."

Ballot transit times are another important potential problem. In their study of UOCAVA voting, the GAO found that transit times for first-class mail can range from as little as five days to as much as a month. A survey by the GAO found that almost two-thirds of all disqualified absentee ballots were rejected because election officials received them after the official deadline. The DoD Inspector General also noted that there are certain difficulties with specific types of mail transit, such as transit to naval vessels underway or to members of the Armed Forces in certain types of service. For example, mail transit averages seven days for 80 percent of mail. However, mail delivery to remote areas and forward deployed locations, such as Bosnia or Kosovo, averaged nine days.

Finally, the Federal Voting Assistance Program's 2004 postelection survey found that problems in ballot transit disenfranchise many voters. Almost one-third of all military personnel and 20 percent of nonfederally employed overseas civilians who did not vote in the 2000 election reported that they did not cast ballots because they either did not receive the ballot they requested or received the ballot late. A related problem encountered at this point is that election officials often find that the Federal Post Card Applications (FPCAs) submitted by voters either have inadequate voting residence address information or contain inadequate or illegible mailing address information. Ballot transit time makes rectifying these problems very difficult.[18]

Equally problematic, the portrait provided by the authors of this report of the existing technology for UOCOVA population is not correct; they assume that the existing UOCAVA voting process was entirely by mail and that threats to that system were generally not of the same scale that might exist where an Internet voting system substituted for the existing process. But UOCAVA voters in many states can use fax technologies for absentee voting. Even states like California, where an electronic voting debate raged before the 2004 election, have allowed UOCAVA voters to use fax technologies for voting. Former California secretary of state Kevin Shelley (one of the strongest foes of electronic voting in the nation before he resigned due to scandals associated with his handling of federal HAVA funds) issued a special directive before the October 2003 gubernatorial recall election allowing UOCAVA voters to use fax transmission for voted absentee ballots in 2003.[19] Unfortunately, election officials typically have not kept detailed historical statistics on UOCAVA ballots, so we do not know how many faxed ballots were cast in recent national elections. But the reality is that UOCAVA voters have the ability to use this very insecure and not-very-private voting method, which is highly vulnerable to many of the same attacks discussed by Jefferson et al. in their report (2004). There have not been any allegations of widespread fraud or irregularities associated with faxed UOCAVA ballots that we have found, but again, given the lack of data available, it is difficult to evaluate fax-based UOCAVA voting.

It is also clear that this report did not recognize how the threats they asserted as being associated with Internet voting are directly analogous to the threats that exist with absentee voting.[20] Denial of service attacks, spoofing, and vote selling and buying occur in absentee voting today, without the Internet (Alvarez 2005). However, because of the small scale of the SERVE project, its spread across a number of states and counties, the procedural realities of UOCAVA voting, and system architecture features designed to mitigate these risks, many of the threats asserted by this report might have been exaggerated. In fact, the risks with Internet voting for UOCAVA voters might have been lower than those with absentee voting, given the very large denial-of-service problem that exists today with paper-based absentee voting among UOCAVA voters.

For example, the ability to launch an attack against the voters using SERVE would have been difficult. The facts of the SERVE project were that it might have involved (at maximum) UOCAVA voters from fifty-one counties in seven states, and only under very optimistic assumptions would there have been as many as 100,000 votes cast using the SERVE system. These voters would have been registered voters distributed across fifty-one counties, but they would have been voting using computers located around the world, many of which were part of well-secured federal networks. Given that each state and county have widely different ballots and ballot styles, mounting an attack where the attacker can successfully replicate the ballot and ballot style of a particular county and get that ballot to the correct voter would have been extremely difficult, if not impossible. Even attempting to mount a systematic attack on the system that was asserted by Jefferson et al. would be difficult.

Furthermore, the authors assumed that were such an attack to occur that it would be difficult to detect and impossible to resolve (while also assuming that attacks in the current UOCAVA voting system can be detected and resolved). The authors ignored the administrative, political, and legal realities of American elections. For example, most election jurisdictions have only small numbers of UOCAVA voters, in many cases literally just a handful of UOCAVA voters. The process of enrolling these voters in SERVE was a two-part *manual* process that involved researching the voters' provenance to ensure they were eligible to participate. Is it unrealistic to assume that election administrators—not to speak of candidates, political parties, the media, and interested citizen observers— would not notice anomalies in UOCAVA balloting in a particular jurisdiction, especially in the hotly contested nature of contemporary American politics. Election administrators have great knowledge of voting patterns in their jurisdictions and would no doubt be the first to notice anomalous activities or results. Politics is a competitive business, and candidates and political parties both are very attentive to the process. It is almost impossible to imagine that competitive politicians and political parties would not notice odd activities or results.[21]

Finally, it is incorrect to assert that nothing can be done about election snafus and anomalous results: these occur frequently, and there are both procedural and legal recourses for situations where odd things happen in elections. If a last-minute denial-of-service attack targeted computers being used for UOCAVA voting from a particular jurisdiction, there is nothing that would prevent a court order from keeping the polls open for additional time so that these voters are not disenfranchised—such court actions occur frequently when polling place voting is disrupted. The most well-known example of this was in 1997, when a judge threw out all of the *paper* absentee ballots cast in the Miami–Dade County mayor's race because of allegations exactly the same as those alleged by Jefferson et al., but perpetrated using the traditional absentee voting process.[22]

In the end, there was quite a bit of heated rhetoric against the SERVE project, but much of the rhetoric was likely unwarranted. First, we note that none of the threats that security experts claim will occur with Internet voting has occurred in the many elections that have tested such systems. Even if we take the criticism of the SERVE project at face value, however, a number of mitigation strategies either were in place or could have been developed to minimize the possible risks and to ensure that any possible problems that arose as a result of the SERVE project could be documented and dealt with. For example, many of the concerns raised about SERVE could have easily been handled through openness and information. One strategy could have been to provide SERVE participants with much "out-of-channel" information (ballot books, briefings in the field, electronic information, list of frequently asked question); this would have given participants a good idea of what to expect when they used the SERVE system and could have helped to foil some of the alleged security problems (like "man-in-the-middle" attacks).

Other strategies could have been to limit participation to a small number of states that were not competitive in the presidential election, for example, Utah or Hawaii; limit the use of the SERVE system to the general election (and not the more complicated primary elections in 2004) and possibly allow only UOCAVA voters to access the federal ballot over the Internet; and to limit access to certain UOCAVA voters from participating states, for example, military personnel using the NIPRNET ("Nonsecure Internet Protocol Router Net," which is used by DoD personnel for secure but unclassified Internet transactions). Many options like these could have provided further assurance to critics that the SERVE system would not allow for widespread manipulation and that any attempted malfeasance would be monitored and stopped.

Although we cannot say whether such options were seriously considered in the internal Department of Defense debates about SERVE after the release of the Jefferson et al. report, we suspect that even with serious risk mitigation strategies the political risks of continuing with the SERVE project might still have been so significant that the project was deemed too controversial to continue. We hope that Internet voting projects like SERVE—projects with strong research and evaluation components—will be initiated again in the future, but in a much less ambitious scope than SERVE, because UOCAVA voters need the sort of increased accessibility and accuracy that a voting system like SERVE can provide and because future trials will give us all a real chance to evaluate the merits of this new technological approach for registration and voting, and instead of relying solely on rhetoric and assertion, we can rely on some actual scientific data.

THE MICHIGAN EXPERIENCE

Ironically, while the SERVE rhetoric was hot, there was an Internet voting election conducted in the United States in January and February of 2004, when the Michigan Democratic Party conducted presidential caucuses in its state. Voters who participated in the caucus could cast their ballots in person on Saturday, February 7, or use by-mail or online absentee voting during a precaucus period starting January 1 and ending at 4 p.m. EST on February 7, 2004. The Michigan Caucus had the largest number of ballots cast over the Internet in any Internet voting trial in the United States, though the relative proportions of ballots cast in person, by mail, or over the Internet were comparable to other recent Internet voting trials. This trial, however, bears the weakness of most Internet voting trials in the United States in that there was no evaluation of the processes and procedures and the outcome of the election. This weakness, unfortunately, does not let us build on the experiences that occurred in Michigan.

The political parties in the United States have broad discretion over the way in which they select their nominees for elected offices. A long-running series of court rulings has established that primaries and caucuses (especially at the presidential level) are viewed as instruments of political parties for their internal candidate nomination procedures and not as procedures necessarily subject to strict governmental regulation.[23] Today, we have a system like that seen in Michigan, where political parties sometimes conduct their presidential preference primaries or caucuses as private elections, while in other situations they will hold them as part of a state-run process. In many states, including Michigan, the political parties also separate the selection process for president from the selection process for state and local offices, which adds complexity to the process.

In Michigan, the Republican Party has usually held a state-run primary in late February or early March. In 1992 and 1996 both parties held state-run primaries, but in 2000 the Democrats attempted to move their candidate selection process to a vote-by-mail procedure before February 12, 2000. Because the Michigan primary would then have been ten days before the New Hampshire primary, in violation of national Democratic Party rules, the Democrats instead held a caucus on Saturday, March 11, 2000. The 2000 caucuses were conducted using absentee voting by mail and through voting at roughly 130 locations throughout the state on March 11. Up to 80 percent of participating voters cast ballots by mail in

2000.[24] For the 2004 election, the Michigan Democrats again decided to utilize a caucus process and to schedule it as close to the New Hampshire primary as possible. In the end they settled on February 7, 2004, as their caucus date.[25] Michigan also planned to allow registered voters to vote using the Internet to cast ballots, in addition to voting by mail or in person. Of the Democratic candidates running for their party's nomination, only Howard Dean and Wesley Clark did not oppose the use of Internet balloting in the Michigan caucus. The other candidates opposed the use of Internet voting, citing concerns about security or inequalities in Internet access.[26] Despite these concerns, the Democratic National Committee (DNC) approved the Michigan plan first in June 2003 and again, after a lengthy appeals process, in November 2003.[27]

The procedures approved by the DNC and used by the Michigan Democratic Party in their 2004 caucus were as follows. Registered voters could apply to receive an absentee or Internet ballot by contacting the Michigan Democratic Party through its Web site, or by mail, fax, or email from January 1, 2004, through January 31, 2004. All absentee ballots (by mail or over the Internet) had to be returned by February 7, 2004. When voters applied through the Michigan Democratic Party to cast a mail or Internet caucus vote, they needed to provide proof of residence, as well as a public declaration that they are a Democrat and that they are or would be a registered voter before the November general election.[28] If the information on the application corresponded with the Michigan secretary of state's Qualified Voter File, the voter would then receive a packet in the mail that included an absentee ballot along with instructions on how to vote over the Internet. No ballots were mailed to addresses other than those on the registered voter list. If the voter was not registered or the information on the application failed to correspond to the registered voter list, the voter could still attempt to cast a caucus ballot at one of more than 500 caucus sites open on February 7. The results from the February 7 caucus were then used to allocate Michigan's delegates to the national Democratic Party convention.[29]

Polling place voting security in the 2004 caucus is similar to what was in place in the 2000 caucus. However, security was heightened in some ways due to the option to vote over the Internet, which required some changes as part of the security environment of the Michigan caucus. Internet security was managed by Election Services Corporation (ESC), a firm that conducts Web-based elections.[30] The system featured two firewalls, redundancy (voters had to verify their choice twice after voting), voter identification numbers, randomly assigned passwords, two kinds of encryption, and two forms of personal identification (city of birth and date of birth). If any of this information was incorrect, voters were not permitted to vote online. Each voter could cast only a single ballot,

TABLE 5.4
How Michigan Caucus Voters Participated

Voting Method	Number	Percentage
By Internet	46,543	28.57
By mail	23,482	14.41
In person	92,904	57.02
Total	162,929	

however, because after the voter uses the voter ID and password, they cannot be used again to cast another Internet vote. The software used by ESC did not require that voters download or install anything on their computer. The ESC election was protected against failure with high redundancy; ESC used more than one communications service provider, all hardware was replicated, and the Web servers and application servers were arranged in "farms" so that, if any piece of equipment failed in a "farm," the others could handle the load until the equipment returned to service. Mail security was also high; the party required two forms of personal identification (city of birth and date of birth) and mailed the ballot only to the address from which the voter had registered. Again, if any of this information was wrong, the voter was not permitted to vote by mail. Mail voters were also required to sign and date their ballots.

Our analysis of the Michigan Democratic caucus begins with the election results. In Table 5.4 we present the breakdown of the total caucus vote by voting method. A total of 162,929 caucus votes were cast and recorded: 29 percent were cast using the Internet, 14 percent by mail, and 57 percent in person on election day.[31] There were 115,245 requests to vote by mail or over the Internet. Of these requests, 40.4 percent resulted in an Internet vote, 20.4 percent in a mailed ballot, and 39.2 percent in no returned ballot.[32] By comparison, in the 2000 Arizona Democratic primary, 42 percent of all ballots were cast over the Internet before election day.[33] The Michigan Democratic caucus produced more votes over the Internet than were produced in Arizona in 2000, but on a percentage basis both Michigan's 2004 trial and Arizona's 2000 trial saw just under a third of participating votes coming from Internet voters. The fact that more Internet votes were cast in Michigan's trial, however, does mean that this was the most extensive trial of Internet voting seen in the United States for what can be considered a public election.

One justification for allowing voting over the Internet was to facilitate a high rate of voter participation.[34] Given the presence of the Internet voting option, Democratic contenders developed different approaches

TABLE 5.5
Participation Rate Comparisons

Election Date	Election Type	Registration	Turnout	Percentage
7 February, 2004	Democratic caucus	6,916,340	162,929	2.40
6 August, 2002	Statewide primary	6,797,293	1,722,869	25.35
8 August, 2000	Statewide primary	6,859,332	1,226,096	17.87
11 March, 2000	Democratic caucus	6,859,332	19,160	0.28
4 August, 1998	Statewide primary	6,300,000	1,375,593	21.83
19 March, 1996	Democratic primary	6,677,079	142,750	2.14
17 March, 1992	Democratic Primary	6,147,083	585,972	9.53

to mobilize their supporters (Seelye 2004). Most of the candidates had links on their Web sites assisting voters to apply for Michigan ballots. The Kerry campaign advertised heavily in college newspapers throughout the state aimed at mobilizing college students. Clarke's campaign generally tried to use the Internet to organize voters in the rural parts of Michigan. Union backers of the Dean and Gephardt campaigns took laptop computers to potential voters and worked to help them apply online to vote. The state party also employed other innovative ways to get voters to make the caucus accessible, both for absentee and election day voters.[35]

In table 5.4 we compare participation in the 2004 Democratic caucus with Democratic primaries or caucuses back to 1992. There are three different types of primaries or caucuses represented in table 5.5: presidential caucuses (2004 and 2000), presidential primaries (1996 and 1992), and statewide partisan primaries (2002, 2000, and 1998). The best comparisons with the 2004 caucus can be found in the other presidential caucuses and primaries. The 2004 caucus had much higher participation than the 2000 caucus and had higher turnout than the 1996 primary. However, turnout in the 1992 primary was almost three and a half times greater than the 2004 caucus. Thus, while the 2004 caucus had greater turnout than two of the previous three primaries or caucuses in Michigan, turnout in the 2004 caucus pales in comparison to past statewide primaries.

In addition to examining turnout, we can also examine the flow of Internet votes in the Michigan caucus over time. In figure 5.1 we reproduce a graph from the Michigan Democratic Party that shows the number of ballots cast on each date over the Internet. This graph shows that despite the lengthy preelection window in which voting was allowed, there is a decided tendency for voters to wait to cast their ballots in the last few days of the election. About one-quarter of the Internet ballots were cast on the last day; almost 70 percent of the Internet ballots were cast in the last four available days.

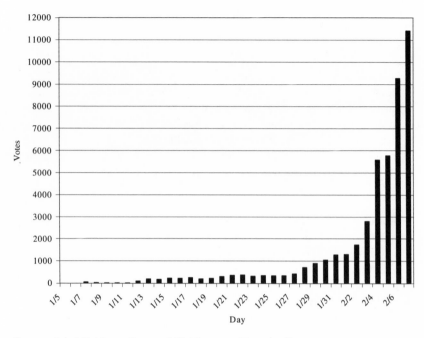

FIGURE 5.1 Michigan Caucus 2004 Internet Votes by Day
Source: Data collected by Betsy Sinclair for the authors

Because we do not have comparable statistics on the flow of mail ballots in the Michigan caucus, we use data from Oregon for comparison. Since 1996, Oregon has used vote-by-mail for their statewide elections. We provide in figure 5.2 graphs that show the daily number of ballots returned in the general elections from 1996 through 2002. Oregon's situation is not exactly comparable, because Oregon provides a much shorter preelection window in which citizens can return their absentee ballots by mail. What is interesting to note between figures 5.1 and 5.2 is the remarkable similarity in the flow of ballots before the election; Oregon's flow of mail ballots looks very similar to that we saw in figure 5.1 for Michigan's 2004 Democratic caucus. Take the 2002 Oregon general election as an example; 28 percent of the 1,276,752 ballots returned in that election were returned on the final day of the preelection period. Also in the 2002 Oregon general election, 62 percent of ballots were returned in the final four days of the preelection period. Recall that in the 2004 Michigan caucus, roughly 25 percent of Internet ballots were cast on the last day and that almost 70 percent of Internet ballots were cast during the last four days of the preelection period.

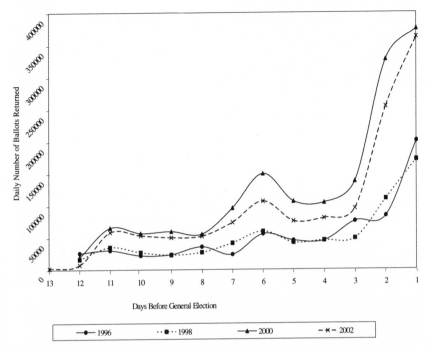

FIGURE 5.2 Daily Ballot Return Rates, Oregon General Elections, 1996–2002
Source: Data collected by Betsy Sinclair for the authors

Given that Internet voting is thought to motivate younger voters to participate in the electoral process, we examine data from the Michigan Democratic Party on the age distributions of those who applied to vote on-line or by mail and of those who voted online in the caucus. Examination of this data indicates that Internet voters were slightly younger than applicants, with the gap between the relative number of applicants to Internet voters slightly larger for the 60 to 69 and 70 to 79 groups, than for the 19 to 29 and 30 to 39 groups. This slight difference is reflected in the average for applicants and Internet voters; the average age of applicants was 51.4, while the average age of Internet voters was 48.5. These data alone make it difficult to know whether the Internet option did indeed draw younger voters to the primary election polls; we do know that in the 2004 general election, 18- to 29-year-old voters made up 20 percent of the estimated electorate, with 26 percent being 30 to 44 years of age, and 54 percent being older than 45 years old.[36]

Finally, we consider how the three different voting methods may have affected the number of votes received by the various candidates. The breakdown of votes per candidate by technology is given in Table 5.6.

TABLE 5.6
Voting Method by Democratic Candidate

	Row Percentages				Column Percentages			
	Internet	Caucus Site	Mail		Internet	Caucus Site	Mail	Total
Mosley Braun	23.93	63.80	12.27		0.08	0.11	0.09	163
Clark	36.84	48.94	14.22		8.7	5.79	6.65	10986
Dean	33.13	49.10	17.77		19.22	14.27	20.42	26994
Edwards	32.15	56.01	11.84		15.14	13.21	11.06	21919
Gephardt	28.60	16.21	55.19		0.58	0.16	2.22	944
Kerry	27.31	57.71	14.97		49.41	52.32	53.7	84214
Kucinich	27.63	62.72	9.65		3.08	3.5	2.13	5183
Lieberman	39.15	21.85	39.00		0.57	0.16	1.13	682
Sharpton	12.24	84.01	3.74		2.97	10.19	1.8	11270
Uncommitted	17.02	48.11	34.87		0.17	0.25	0.71	476
Write In	38.78	38.78	22.45		0.08	0.04	0.09	98
Totals	46,543	92,904	23,482		46,543	92,9.4	23,482	162929

Note that all candidates except for Gephardt and Lieberman won more caucus votes than Internet or mail votes. This is, in all likelihood, an artifact of Gephardt and Lieberman having dropped out of the caucus by February 7. Kerry had the most votes from each method of technology. The role of Internet voting does not appear to have changed the election outcome. The preferences of Internet voters seem to be almost identical to those of polling place voters. This suggests that claims that the Internet created a form of discriminatory access are not robust, as the preferences of polling place voters are similar to Internet voters.

There are several interesting lessons to draw from the Michigan Internet voting trial. First, contrary to the concerns of critics, there were no successful attacks from pranksters and hackers. One argument used by critics was that this primary was not important, and the primary was not worth hacking.[37] Of course, the candidates who lost the election—especially Howard Dean, who supported the Internet voting trial—would probably dispute that the Michigan caucus was not important. Second, concerns about security did not seem to bother voters, who cast more Internet votes than mail ballots (and half as many Internet votes as polling place votes). This is especially surprising given that not every voter has easy access to a computer, although public computers (such as those in public libraries) were made available and the number of polling places was increased to ensure all voters had access to caucus voting. In a survey conducted by CBS News of 479 Michigan Internet voters, 67 percent said they used Internet balloting for the convenience. Of these voters, 90 percent reported using a home computer to cast their ballot, while 8 percent used a computer at work (Moses 2004).

Third, when we compare the problems that Internet voters and polling place voters encountered, it is evident that the problems associated with polling place voting can equally be, if not more, pernicious. When polling places move on election day, it can be difficult for voters to vote. Moreover, when voters encounter problems, they are able to cast a provisional ballot, but it is not always the case that the ballot will be counted or counted in total. Although critics argued that Internet voting would be a cause of disenfranchisement, it was polling places in Detroit that seem to have disenfranchised voters.

Finally, we note that many campaigns failed to develop strategies to maximize the turnout of their potential voters by leveraging Internet voting. Any could have used a strategy similar to that used by the unions and, for example, stationed campaign staff with laptops equipped with wireless Internet connectivity outside of churches in Detroit on consecutive Sundays. On the first Sunday, the campaign staff would register voters and the following Sundays the staff would allow voters to cast ballots using the laptops. In essence, the campaigns would bring the voting booth to the voter.

INTERNET VOTING: WHERE DO WE STAND TODAY?

There are many ironies in the debate over Internet voting. First, while the United States continues to debate the efficacy of this form of voting, other nations are forging ahead with experiments and efforts to allow their citizenry to use this medium. Successful Internet voting experiments are conducted every year, but the debate in the United States remains stuck. We continue to study and gather information on trials in other nations like Estonia, Switzerland, and the United Kingdom, where innovative efforts to experiment and implement voting procedures using the Internet and other new electronic technologies are ongoing. Among these other technologies are trials of voting using "short message service" or "instant messaging" in Switzerland and the United Kingdom. Estonia has provided all citizens with a digital signature that can be used to authenticate a person in a wide array of electronic government transactions, including voting. By not experimenting with these new technologies, the United States could fall behind the rest of the world.

Second, the problems that led to the creation of SERVE in the first place—the difficulties associated with UOCAVA voters casting ballots by mail—remain and in many ways are exacerbated by the ongoing deployments of large number of military forces and associated civilians overseas. In the 2000 election debacle in Florida, issues associated with ballots coming from overseas voters figured prominently in the debate about who won that election—and how to improve the process of election administration there and in the rest of the nation. But since that election, and as part of the reframing of the election reform debate, the discussion has shifted from a frame of accessibility and accuracy to one concerning security. In such an environment, it is difficult to maintain much interest in UOCAVA voters and the difficulties they face when they try to participate in our democratic process.

Third, with Internet voting off the table, local election officials and UOCAVA voters are forced to look to less secure alternatives (in addition to the conventional paper-based by-mail process), especially email and fax. In the run-up to the 2006 elections, the media "discovered" once again that the by-mail process does not work for UOCAVA voters and calls were made for the development of new voting mechanisms to serve these voters (Jelinek 2006). Unfortunately, until new approaches are developed, tested, and implemented, UOCAVA voters are potentially much more likely to be disenfranchised when they vote. They will continue to lack access to the new technologies available to voters in traditional precinct or early voting situations that allow for ballots to be checked for common errors. Or they will be forced to rely upon possibly less secure and less private technologies, like fax and email, if they do want to try to get their ballot and return it relatively quickly. Until we can make some technological progress, these voters may remain among the least represented in our political process.

The ultimate irony is that Europe is the region of the world where the precautionary principle is most often espoused as the way in which to address risk but is also the region that is most active in experimenting with electronic voting. Critics of electronic voting likely wonder, given their sharp concerns about risk in other policy areas, why the Europeans do not take their votes and their democracy more seriously and protect it from these threats. It may be that they do take risks seriously but, weighing the costs against the benefits and in keeping with the Enlightenment's promotion of science over raw faith, they see the benefits to conducting experiments with this new technology. Given the use of such limited experiments, the scientists may have decided that the costs of failure have been minimized and the scientific knowledge that is gained can lead to a better and more inclusive democracy.

Chapter 6

THE PERFORMANCE OF THE MACHINES

In 2004 there was one primary election where voting fraud was alleged to have altered the outcomes of an election. In Texas, Democrat representative Ciro Rodriguez alleged that fraud in a recount caused his 145-vote victory in a primary to turn into a 203-vote defeat (Mock 2004). The irregularities? First, the vote total in the challenger's hometown exceeded by 115 votes the total number of voters who were recorded as having cast ballots. Second, 304 uncounted ballots were "found" after the election and three-fourths were cast for the challenger. Were these lost and found ballots electronic? Actually, they were paper-based optical scan ballots. Given the intense media focus on electronic voting in 2004 and not on paper-based voting, this story slipped under the radar of most of the mainstream media.

As we have noted repeatedly, electronic voting was under an intense microscope in 2004 and again in 2006. Interest groups, the media, and other election watchers were carefully examining how this voting technology performed, especially in places where e-voting was being rolled out for the first time. The preelection rhetoric and media coverage before both of these elections suggested that the machines were going to be disastrous, with the machines vulnerable to malfunctions, problematic implementation by poll workers, and fraud. When the 2004 election was completed, however, there were only two locations where problems arose with electronic ballots, and both problems were on older electronic voting technologies; following the 2006 midterm election there was one location with a significant problem involving electronic voting technologies. There were also counties with significant problems with paper ballots, but such problems rarely were covered in major national media. The reality, that both paper and electronic voting systems generally performed reliably in most places, did not stop critics of electronic voting from attempting to discern fraud in the system, and the postelection rhetoric—much of which was distributed in a relatively unfiltered and unsubstantiated way on the Internet—was stoked with claims that electronic voting had led to President Bush receiving more votes than was "normal" in a presidential election. Instead, the real story of the 2004 election was that John Kerry could not muster the necessary votes to beat an incumbent president in wartime, in a context where fears of terrorism and moral decline may have lead many voters to feel the need to stick with the president who, while flawed, was a known quantity. In the end, Kerry lost the popular vote only by about 2.5 percent.

In this chapter, we examine the performance of electronic voting since 2000 from several vantage points. First, we begin by providing a baseline for evaluating electronic voting, and that baseline is the alternative: paper ballots that are either electronically or handcounted. Examining the performance of paper ballots in the 2000 election, as well as examining the implementation of paper balloting in elections through 2005, allows us to study electronic voting in an appropriate context. Second, we look at the several specific cases where problems were alleged to have occurred with electronic voting systems since 2000 to determine whether these problems are unique to electronic voting or whether they are similar in size and impact to problems that occur with paper ballots.

Third, we turn our attention to examining how electronic voting affects residual vote rates when compared to other voting technologies. We are specifically interested in examining the number of ballots that go uncounted for any given race on the ballot across machine types—across counties and across precincts within specific counties. This analysis informs us about the impact of voting technology on residual votes across demographic groups and the impact of switching from one type of voting equipment to another. Finally, we consider how specific groups who should benefit from electronic voting, especially the disabled, fared in the transition to the new system.

THE DIFFICULTY OF VOTING ON PAPER

In the run up to the 2004 election, one scholar who studies election reform told the following story.

> Imagine that a Martian landed on Earth and was interested in learning about election reform but could only learn about it by reading current newspapers. The Martian could easily be forgiven for thinking that the reason people are so interested in election procedures and voting technology today is because in the 2000 election someone hacked the electronic voting machines in Palm Beach County and Pat Buchanan was elected President.[1]

As we all know, the reason there is so much interest in studying election procedures and voting technology today is because of a "perfect storm" in the 2000 presidential election, with this storm centered primarily on Florida. In that state, a very tight presidential election, poor election administration practices, problematic paper ballot voting systems, and a state election code that did not have adequate provisions for recounting paper ballots, led to a system failure and produced one of the biggest election crises since 1876. Fortunately, there were several studies of the paper ballots in Florida conducted after the election, and other states like Georgia examined the performance of their voting technologies after 2000. Georgia is an important case because it used every voting technology except electronic voting in the 2000 election. There have also been studies by political scientists of voting outcomes nationally by voting technology that are critical to understanding the current debate over electronic voting.

The failure of paper ballots in Florida was documented by work conducted by a consortium of media organizations, which hired the National Opinion Research Center (NORC) to study the status of uncounted ballots in Florida in 2000. Residents of Florida voted on five platforms: Votomatic (prescored punch cards that produce chad), Datavote (punch cards that are scored with a machine by the voter), machine-counted optical scan ballots, lever machines, and paper ballots. Of the first three voting methods, the two punch-card ballot types had residual vote rates of 3.8 and 3.7 percent, respectively. Together, these two punch card technologies produced almost 54,000 undervotes and more than 89,000 overvotes. The Votomatic was more likely to produce undervotes, and the Datavote was more likely to produce overvotes. These two punch card voting technologies produced more than 143,000 uncounted ballots. Optical scan balloting produced a residual vote rate that was lower—1.3 percent—but this still totaled 31,775 uncountable ballots in an election decided by 537 votes (Wolter et al. 2003, 5).

As we noted previously, voting on a paper medium is a two-step process: voters render their candidate and issue preferences by marking the ballot with specific choices;[2] and then election officials discern these preferences. Because of the secret ballot, voters can never know if the election officials discerned their preferences accurately on a paper ballot. The NORC study has two interesting findings for appreciating the issues associated with counting votes on paper. The first issue we consider is reliability: do individuals evaluating a voted ballot determine the same indications of voter intent? We can answer part of this question because the NORC study reviewed ballots using teams of three coders. In three counties, all three coders reviewed all ballots for under- and overvotes. In the remaining counties, three coders reviewed all undervotes but only one coder reviewed all overvotes. For each ballot type, NORC produced a set of reliability statistics, assessing "the proportion of agreement among all pairwise comparisons between coders" (Wolter et al. 2003, 5). With ten presidential candidates and three coders, each ballot review consisted of thirty pairwise comparisons. A second set of reliability statistics examined all three-way comparisons, which total ten per ballot (one per candidate). These reliability statistics can range from 0 to 1; an agreement score of 1 means there is perfect agreement among the coders.

Examining only the undervoted ballots, for all nonabsentee ballots the pairwise agreement statistics for Bush and for Gore are 0.90 and 0.88 respectively. These scores varied by voting machine type. For Votomatic, the pairwise agreement statistics for nonabsentee ballots were 0.89 and 0.89 for Bush and Gore, respectively, and were 0.98 and 0.94 for ballots from the Datavote system. Given that punch cards are being discontinued as a consequence of state and federal election reforms, the most interesting statistics are those for optical scan ballots. On the optical scan platform, the pairwise agreement statistics for Bush and Gore are 0.96 and 0.93, respectively, for nonabsentee ballots, and 0.97 for both candidates for absentee ballots. Note that the agreement for Datavote and for optical scan technologies is very similar. This means that, for every 100 ballots cast on optical scan ballots, there will be three cases—3 percent of votes cast—where the votes on the ballot are subject to differing interpretation. Also, note that there are differences between the agreements that ballot counters have for Bush compared to those for Gore. The difference for optical scan is larger than the difference in the 2000 election outcome between Bush and Gore. Examining only the overvoted ballots in the three counties where teams of three examined the ballots, the pairwise agreement statistics are 0.97 for both candidates on Votomatic ballots, 1.00 for both candidates on Datavote, and 0.99 for both candidates on optical scan ballots. Again, in 1 percent of cases, the votes cast on optical scan ballots will be subject to conflicting interpretations among trained individuals.

Second, we consider whether different means of counting paper ballots produce different outcomes. The study found new votes for both candidates, based on nine different standards for counting ballots. Using any of the nine counting standards resulted in a change in the vote totals from the official results. There were swings of more than 100 votes under seven of the nine standards; six of the nine standards have vote swings that produce a Gore victory. Ironically, many of the outcomes that lead to a Gore victory were produced by recounting procedures that were not being promoted by Gore and his legal team in the aftermath of the Florida 2000 presidential election. These findings also call into question the comment made by advocates of paper ballots that "we can just recount the ballots." In Florida, recounting the ballots could result in nine different vote totals and with six resulting in one candidate winning and three resulting in the other candidate winning.

What these findings show is that counting paper ballots is a more difficult task than is generally acknowledged. Even when ballots are hand-counted, the likelihood that three individuals will agree on the voter's intent, while high, is not 100 percent. The variations in counting optical scan ballots ranged from 1 percent for overvoted ballots to 3 percent for all ballots. These small differences can change the outcome of any given election. In addition, the NORC study shows that paper ballots, whether an optical scan or punch card style, are highly susceptible to the standard used to determine whether a ballot should be counted and for the determination of voter intent. Different standards can produce vastly different outcomes; in the NORC analysis, the shift in the vote count could have been as much 708 votes—and under many of these scenarios, we would have had a President Gore in January 2000, not President Bush.

These findings about the difficulty of counting paper ballots are reinforced by a study of recounts in New Hampshire conducted by Stephen Ansolabehere and Andrew Reeves (2004). In this study, they examined the percentage of ballots cast that differed between the initial count in an election and in the recount. For hand-counted ballots, the weighted average difference between the initial count and the recount was 2.49, although when one town was excluded, the weighted average declined to 0.87. However, the error rate between the initial count and the recount ranged up to 21.88 percent with all localities included, and 4.08 percent with the one outlier town removed. For optical scan ballots, the weighted average was lower than with hand-counted ballots—only 0.56 percent—but again, the range extended all the way to 8.27 percent. Ansolabehere and Reeves (2004, 6) discuss the practical effect of this counting difference by giving the example of "an election with 10,000 votes where the candidate in question received exactly half of the votes. The predicted discrepancy between counts is 1.24 percent if the town tabulates by hand and .71 percent if the town uses an optical scanner."

Here it is helpful to imagine the application of the standards of the precautionary principle not to electronic voting but to paper-based voting systems. We noted earlier that the precautionary principle is based on the idea that the decision to mitigate a potential risk should not require the existence of absolute proof that it will come to fruition. The 2000 election clearly showed that the risks associated with paper ballots are not potential. In fact, the aftermath of the 2000 election and the subsequent studies of the problems associated with punch cards—and somewhat less so with optical scan voting—strongly suggest that the risks associated with such systems are high. After all, getting three people to agree what constitutes a vote on these systems was shown to be less than 100 percent and varying standards of what constituted a vote produced different results, most of which would have changed the outcome of the election. The technology created a great controversy about the conduct of elections and lowered public confidence in the electoral process and the conduct of American democracy.[3] To press this point to its extreme in order to make a point, many of the policy issues that America faces in the early twenty-first century can be traced back to the use of paper ballots in Florida and the election of George W. Bush that resulted from the use of this technology.

OTHER ISSUES WITH PAPER BALLOTS

It is also important to understand the "snagging" process involved in the counting of paper ballots. We bring up this process not to be conspiratorial—although we have no doubts that critics of electronic voting would excoriate the practice as an obvious point for introducing fraud into an election if it was associated with electronic voting—but to note again the difficulty associated with divining "voter intent" from a paper ballot. Snagging is a standard procedure in election administration designed to identify problem ballots before they are fed to tabulation devices—ballots that are torn, bent, or not clearly marked—and then to remake (when possible) those ballots so the tabulation device can more readily discern the voter's intention.

In the 2005 Los Angeles City mayor's race, this process became controversial when it was revealed that the city clerk had "inspected every ballot by and [overmarked] ballots where the voter's ink mark might not have been read by vote counting machines. . . . The clerk said in his report [to the city council] that he wanted to make sure every vote was counted, so he had election workers inspect each ballot and use a light blue pen to mark the ovals where voters' ink marks might have been too small to be read by the machine" (Mc Greevy 2005). The city clerk (who in the organizational chart of the city was an "employee" of the existing mayor who was running for reelection in 2005) was accused of creating the perception of improprieties and ballot tampering. An adviser to one of the candidates noted that the process had "a great deal of subjectivity," although there was "nothing to indicate voter fraud" (Barrett 2005, Olov 2005). The election controversy grew when it was reported that the software used to count the ballots was changed before the election without state approval (Anderson and Barrett 2005). Although there was no controversy over the actual outcome of the election, it is possible to see how paper ballots are subject to change by election officials through a process that can be questioned as being subjective and improper. Note that this also means that a voter's ballot that is counted in many cases will not be the ballot that the voter marked. Instead, the voter's ballot has been interpreted by a third party, who is assuming what the voter meant to do and remarking the ballot to reflect this interpretation.

Finally, we note that it is not as though fraud with paper ballots is a thing of the past. Dan Tokaji (2004) notes that in the case *Weber v. Shelley*, which challenged the use of DRE voting machines in Riverside, California, that did not have a voter-verified paper audit trial (VVPAT) component, the court explicitly noted that critics of DREs have failed to show that "the paperless DRE was 'inherently less accurate, or produces a vote count that is inherently less verifiable, than other systems' What is missing, in other words, is any evidence showing that DREs are *comparably* less accurate or reliable than other systems" (1751). Tokaji then gives an array of example from the past and present:

> In New Mexico's 2000 presidential vote, some 252 early-voting ballots were reported missing and another 1,300 to 1,600 "damaged votes" were rejected because of stray marks or other problems. During a 2000 election in Benton County, Arkansas, a ballot box was "misplaced," and only to reappear after some 12 hours with its label peeled off and the box wet from sitting out in the overnight rain. In a 2002 Illinois assembly election, ballots cast in one of the precincts could not be located at all, causing the trial judge to order a new election. (1752)

It is critical to remember that the hacking that can be done to an electronic voting system can also be done to a paper-based system, especially a hack that would target the electronic tabulation device used to count the paper ballots. Critics may argue that with paper ballots any such hack can be overcome, because you can just count the ballots again, but in many states this is simply not the case. States often have very strict rules governing when recounts can occur; if the hack generates an outcome that is outside of the bounds of the recount law, then it may not be possible to simply "recount" the ballots. Also, if the hack is done well, then there may never be sufficient postelection evidence to motivate anyone to seek a recount of the paper ballots, even if allowed under state law. As we have noted earlier, paper ballots can be recounted repeatedly, and depending on the standards and whims of those recounting the ballots, there can be decidedly different election outcomes in a close race. Clearly, we need valid methods for recounting ballots and auditing voting systems, whether they are predominantly electronic or paper.

Civil Rights and Voting on Paper

A second critical issue associated with paper ballots is whether these voting technologies are biased against certain populations of voters. In the aftermath of the 2000 election, the issue of racial bias in voting systems came to the fore, and the volume of research on this issue accelerated rapidly. Michael Tomz and Robert van Houweling (2003) published one of the key studies in this area and summarize this literature. Critically, they note that "a growing body of evidence suggests that blacks cast invalid ballots more often than whites" (47). Studies of precinct-level data from Palm Beach, Broward, Duval, and Miami-Date counties in Florida, cross-county data in Florida, the NORC data for Florida discussed previously, studies of voting in California and Georgia, and multistate studies all have found a strong relationship between high percentages of rejected ballots and high percentages of black voters in a precinct or jurisdiction.[4]

Tomz and van Houweling (2003) found that there was a 4 percent race gap in precincts that used punch cards and a 6 percent gap in precincts that used optical scan balloting. By contrast, the gap in precincts using DREs was 0.3 percent, which is ten times less than the punch card gap. They noted that precinct counting can lessen the racial gap problems associated with optical scan balloting, but this finding did not hold true across all precincts studied. Another study by Stephen Knack and Martha Kropf (2003) of voting in 1996 found that precinct count optical scan voting systems minimized racial gaps.

The key issue under HAVA, of course, is the ability of a voting technology to accommodate the needs of voters with disabilities. Specifically, Section 301 states that a HAVA-compliant voting system must "be accessible for individuals with disabilities, including nonvisual accessibility for the blind and visually impaired, in a manner that provides the same opportunity for access and participation as for other voters." The problems that voters with disabilities encounter voting on a paper-based system are well-known. These voters cannot vote independently on paper, cannot know that their vote is being cast in keeping with their preference, and cannot know that they voted the entire ballot.[5] What we do not know, and what we do not have much research on, is how the wide variety of voting systems designed to allow voters with disabilities to cast a secret and private ballot actually perform—how accurate the different assisted-voting voting systems are, how reliable they are, nor how usable they are.

BENEFITS OF SWITCHING TO DREs FOR MINORITIES AND VOTERS WITH DISABILITIES

One key issue associated with voting systems is to determine which systems lose the fewest votes—that is, have the lowest residual vote rates—and which systems accomplish this goal in a way that minimizes any racial disparities. Two studies by Charles Stewart of MIT, conducted as a part of his role in the Caltech/MIT Voting Technology Project, can help to inform our understanding of this issue, especially as related to moving from paper-based systems to electronic voting systems. His first study examines the transition to electronic voting in Georgia in 2002, and a second examined voting nationally in 2004.

Georgia is an excellent case for understanding the impact of moving to electronic voting. With 159 counties, the state had a broad range of voting technology usage in 2000: 58 percent of votes in 2000 were cast on punch cards, 31 percent were cast on lever machines, 7 percent were cast on hand-counted paper ballots, 3 percent were cast on optical scanned ballots, and a residual 1 percent were cast on an undetermined type of machines (Stewart 2004). In 2002 the state moved to an electronic voting platform statewide, adopting the Diebold Accuvote-TS system.[6] To appreciate why the state made this transition so quickly after the 2000 election, it is critical to review the residual vote rates in the state, by voting technology, for the 1998 gubernatorial election and the 2000 presidential election.

As Charles Bullock and M.V. Hood (2002) found in their study of residual voting in Georgia, counties with lower education levels, high percentages of African American voters, and high numbers of new registrants were more likely to have high residual vote rates in the 2000 election. Stewart (2004) examines the 1998, 2000, and 2002 elections. He finds that, in 1998, punch card voting *outperformed* all other voting methods on an absolute level for residual vote rates. After controlling for population and wealth differences between counties, optical scanning has a marginal benefit in reducing residual vote rates by 0.5 percent. In the 2000 presidential election, the residual voting rate for punch cards was almost twice the 1998 rate—4.7 percent in 2000 compared to 2.4 percent in 1998—and in this case the residual vote rates for all other technologies were lower than the punch card rate. However, the residual vote rates for all technologies were almost twice as high in Georgia compared to other voting technologies in 2000. It is also helpful to compare residual vote rates for statewide offices in 2002 to the rates in 1998 to determine what the overall impact was of shifting to the Diebold Accuvote-TS. For every statewide race, the residual vote rate declined. The gains ranged from 1.7 percent in the governor's race to 7.5 percent in the race for agriculture commissioner. In one county, the residual vote rate was 20 percent in 1998 and declined to 0.8 percent in 2002.

Stewart (2005) also examines what happened when counties switched voting technologies between 2000 and 2004. He found that there were four major types of equipment changes—punch card to DRE, optical scan to DRE, lever machine to DRE, and punch card to optical scan—with a large residual number of counties and voters casting ballots on the same technology in both elections. In counties where voters used the same technology in both 2000 and 2004, there was a decline in residual votes of 0.61 percent between the two elections, suggesting that election reform efforts and voter education provided some reduction in residual votes irrespective of any technology improvements. However, switching technologies did have marked benefits in reducing residual vote rates. In counties that switched from punch cards to optical scan or DRE technology, the residual vote rates declined by 1.12 and 1.46 percent, respectively. Likewise, switching from optical scan to DREs reduced residual vote rates by 1.26 percent. This analysis also found that engaging in a statewide reform of voting technology as was done in Georgia was the most effective method for reducing residual votes. Perhaps one of the most interesting findings is that counties that switch from optical scan to DRE make particular gains in residual vote reductions, which is impressive in part because optical scanning systems typically have low residual vote rates.

By looking to Florida, just south of Georgia, we can see the benefits of electronic voting for voters with disabilities. There is strong anecdotal evidence regarding the benefits that have accrued to voters who have either limited visual acuity or limited language proficiency. An example of the cost of not having an electronic voting system appeared in the *St. Petersburg Times* after the 2004 election:

> David Bearden says he was heckled as he left his Brooksville polling place Election Day because he hadn't voted Republican—and a line of people knew it. When Susan Cook tried to cast her ballot, a stranger approached her and started talking to her. Cook said she must not have looked like she was voting because people were listening to her repeat her choices. Hernando County has no voting machines accessible to the visually impaired. And dozens of blind voters who went to polling places last week had to speak their choices aloud to poll workers who filled in ballots to be scanned by machines. . . . Bearden and Cook say their right to a secret ballot was violated because other voters waiting in line at small polling places heard their choices for each race. "It was atrocious," said Cook, who voted at New Covenant Baptist Church in Brooksville. "What's the concept of a secret ballot if everybody knows how you voted?" Bearden filed a verbal complaint with the state Division of Elections. He said a woman teased him about his vote as he and his guide dog left the polling place at Hillside Community Church, east of Brooksville (Liberto 2004).

THE OVERALL PERFORMANCE OF DREs IN 2004

The key question regarding DREs in 2004 was, How well did they perform? We can answer this by looking at data collected nationally on system performance or by examining specific cases where problems emerged to determine their cause. Before delving into these data in this and the following sections, a note about these data is in order. On several occasions, critics of electronic voting have made the claim that, because electronic voting machines are insecure, using data on residual votes is not a meaningful metric because the data from the electronic voting machines could have been manipulated. However, because all voting data are subject to potential manipulation and residual votes were one of the key issues in the 2000 election in Florida, we focus on this metric in our analysis that follows. We can analyze only the data and information that are available to researchers, and we have no reason to believe that the reported data have been manipulated for any particular voting technology.

The first question we examine is a basic one: what was the residual vote rate—the number of ballots uncounted because of overvotes, undervotes, or other problems—for DREs compared to paper-based systems, and how does this rate compare from 2000 to 2004? Table 6.1 presents the residual vote rates for all voting systems for 2000 and 2004, using data by both counties and counties weighted for turnout in the counties. For all voting systems except paper ballots, the residual vote rates declined from 2000 to 2004. The best-performing equipment in 2004 was the lever machine, followed by electronic and optical scan technologies, which performed almost identically.[7] The overall reduction of residual votes across voting technologies strongly suggests that there were environmental changes that reduced the overall residual vote rate across voting platforms. Voters, election officials, and poll workers were all much more cognizant of the potential for problems with their ballots, and there were many changes in process and voting technologies, all of which likely helped to reduce residual votes across all voting systems.

The data in Table 6.1 also allow us to compare the 2002 and 2004 gubernatorial and senatorial elections. These results have to be considered carefully, because the number of states that have gubernatorial elections in presidential election years is much smaller than the number of gubernatorial elections held in nonpresidential election years. Likewise, in many states the mix of voting technologies used in these states changed between 2002 and 2004. With those caveats in mind, note that for senatorial elections, residual vote rates between 2002 and 2004 declined for optical scan and electronic voting to approximately the same rate, while the residual vote rates for punch cards and lever machines climbed. This result could be the result of better-managed and more innovative counties switching from technologies discouraged under HAVA—punch cards and lever machines—to better-performing voting technologies. Early adopters are often "champions" of reform, which leaves more cautious jurisdictions using the older voting systems.

TABLE 6.1
Average Residual Vote by Machine Type and Year in U.S. Counties

| | Presidential Elections | | | |
| | Counties[a] | | Voters[b] | |
Machine type	2000	2004	2000	2004
Punch card	2.6	2.0	2.6	1.6
Lever machine	2.2	1.1	1.9	0.9
Paper	2.2	2.2	1.9	2.1
Optical scan	2.1	1.4	1.3	1.1
Electronic	2.4	1.6	1.8	1.0
Total	2.3	1.6	1.9	1.1
	Gubernatorial Elections			
	Counties[a]		Voters[b]	
Machine type	2002	2004	2002	2004
Punch card	2.5	3.8	3.5	3.9
Lever machine	2.9	4.5	2.9	5.4
Paper	1.6	2.1	1.8	2.3
Optical scan	1.7	2.3	1.9	2.1
Electronic	1.9	3.2	1.3	2.9
Total	2.0	2.4	2.4	2.8
	Senatorial Elections			
	Counties[a]		Voters[b]	
Machine type	2002	2004	2002	2004
Punch card	2.8	5.7	2.4	5.2
Lever machine	9.6	10.2	6.0	10.2
Paper	3.7	3.6	5.4	4.5
Optical scan	3.9	3.2	3.5	3.5
Electronic	5.4	3.3	4.0	3.6
Total	4.4	4.0	3.4	3.6

[a] Data for counties represents unweighted county residual vote rates

[b] Data for voters represents county residual vote rates weighted by registered voters

This last finding leads to a second question to examine here: what is the affect of switching systems on residual vote rates? Table 6.2 shows that there are numerous combinations of system switches: a county could switch *from* paper ballots, punch cards, lever machines, optical scan ballots, or electronic voting technologies *to* either optical scan equipment or electronic voting technologies. When we examine the reduction in residual

TABLE 6.2
Average Change in Residual Vote Rate for Counties Switching Technology, 2000–2004

Equipment Used in 2000	Equipment Used in 2004	Change in Residual Vote Rate(%)	Number of Counties/Voters in 2004
Punch card	DRE	−1.30	48/6.8 million
Optical scan	DRE	−1.50	36/2.6 million
Lever machine	DRE	−0.80	29/145,000
Punch card	Optical scan	−2.00	107/11.4 million
All other equipment changes		−1.30	26/273,000
Same equipment in 2000 and 2004		−0.30	1830/51.7 million

votes from 2000 to 2004 for counties that switched equipment, we find that in all cases the switch to electronic voting reduced residual vote rates. So, for example, switching from optical scan ballot equipment to electronic voting equipment reduced residual vote rates by 1.5 percentage points. Switching from punch cards to electronic voting equipment reduced residual vote rates on average by 1.3 percentage points and switching from punch cards to optical scan reduced the residual vote rate by 2.0 percentage points. Approximately 9.6 million voters switched to electronic equipment by 2004, and 11.4 million switched to optical scan equipment. Over 50 million voters used the same equipment in both elections.

The information in Table 6.2 gives us a clue that switching voting technology resulted in sizable reductions in the residual vote rate. However, we need to control for other factors in order to ensure that these results are meaningful. For example, it could be that some intervening factor, like county size, was actually critical for understanding why we see these reductions in residual vote rates. In the appendix to this chapter, we show the results of an advanced statistical model that estimates the benefits of switching from punch cards to another voting technology. For ease of comparison, we illustrate the benefits of switching graphically in Figure 6.1. Here, we present the estimated percentage change in the residual vote rate for a county changing from punch cards to other voting technologies (Halvarsen and Palmguist 1980; van Ganderan, Kees, and Shat 2002; Kennedy 1981). The lower the bars go, the greater the estimated reduction in residual votes that come from switching to the new voting technology from punch cards, which are the poorest performing voting technology in the analysis. There is a reduction of residual votes of just under 30 percent when moving to electronic voting, just under 35 percent when moving to optical scan ballots, just

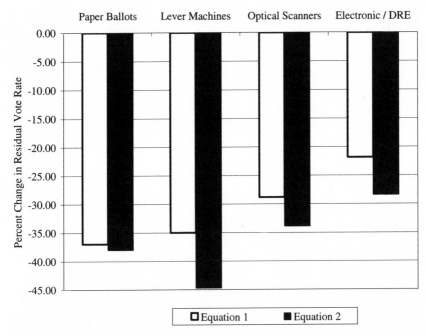

FIGURE 6.1 Effects of Technology Change, Switching from Punch Cards
Source: Data from the Voting Technology Project: analyzed by Delia Bailey

under 40 percent when moving to hand-counted paper ballots, and just under 45 percent when moving to lever machines. These figures are slightly lower when we do not control for turnout; the dark bars illustrate the reduction when we control for turnout by county.

The information in Figure 6.1 is based on estimating the reduction in residual votes when a switch from punch cards occurs. These estimates are not exact; we have provided the best estimate, but there is a probability that the estimate could fall elsewhere along a range of potential outcomes. When we consider the range of outcomes for switching from punch cards to each voting technology, we see that the results vary across voting platforms. For example, although hand-counted paper ballots perform best in the analysis, there is also a very wide range of possible outcomes, from a high of a 60 percent reduction in residual votes to a low below 20 percent. The range of outcomes for paper ballots encompasses that for the other voting systems, so we cannot be completely confident that hand-counted paper ballots would outperform other selected voting technologies. If we examine optical scan ballots and electronic voting, we see that electronic voting has a broader range—from approximately a 15 percent reduction to a 40 percent reduction—that encompasses almost the entire range of possible reductions in residual votes for optical scan voting. We interpret this to mean that switching to optical scan produces a high level of performance within a relatively narrower band of possible outcomes compared to electronic voting. The overlap in probabilities also means that many localities that switch to electronic voting will equal or exceed the performance of all but a few optical scan counties. However, the poorly performing electronic voting counties fall below the poor performing optical scan counties.

INCIDENT REPORTS OF ELECTION PROBLEMS IN 2004 AND 2006

There have been a number of efforts to collect systematic data on problems with voting equipment, and in this section we examine two of those efforts. First, we look at the effort to gather national-level information on problems with voting equipment in 2004, data released by the U.S. Election Assistance Commission (EAC), as part of its 2004 Election Day Survey.[8] Two components of its Election Day Survey were collecting data from each election jurisdiction in the United States on the voting system it used in the 2004 presidential election and asking state election directors to identify by county and precinct where any of a list of voting system malfunctions occurred: power failure, broken counter, computer failure, printer failure, screen failure, fatal damage to the machine, modem failure, scanner failure, ballot encoder or activator failure, audio ballot failure, or other voting machine malfunction. Unfortunately, the EAC survey team was unable to achieve anything near a reasonable response rate for the questions on voting system malfunctions, as only twenty states provided detailed data in response to these questions (twenty-one states provided no information, two states stated that no voting system malfunctions occurred). In the end, the report has data from only 485 of the 6,567 election jurisdictions in the survey database. Of the jurisdictions that had data on voting machine malfunctions, 43 percent (210) used optical scans, 39 percent (191) used electronic voting equipment, and the remainder used some other type of voting system.

In Figure 6.2 we graph the reported malfunction data, ordered by the total incidence rate of each problem across all voting systems, and then just for jurisdictions using electronic voting systems and optical scans. Overall, in the jurisdictions reporting voting system malfunction data, the most prevalent problems reported were scanner failures (almost 23 percent of reported problems), printer failures (20 percent), computer failure (15 percent) and screen failures (14 percent). Not too surprisingly, when we consider the differences between optical scans and electronic

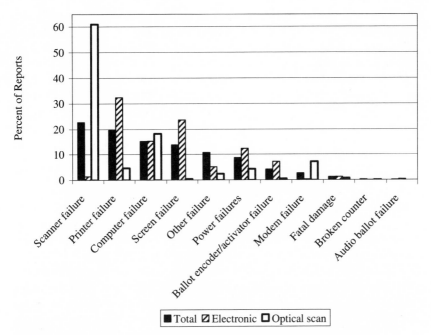

FIGURE 6.2 Election Assistance Commission, Reports of Voting Machine Malfunctions

voting machines, we see that the scanner failure problem accounted for 61 percent of the reported problems with optical scanning devices, with the second-most reported failure of those voting machines being computer failures (18 percent)

As for electronic voting machines, printer failures accounted for about a third (32 percent) of the reported malfunctions. The other high-incidence malfunctions reported for electronic voting systems were screen failures (24 percent), computer failures (15 percent), and power failures (12 percent). Among the reported electronic voting system malfunction data presented here, the most prevalent reported failures regard printer failure, not necessarily failures of the sorts that foes of electronic voting technologies have typically been concerned with.

We need again to be cautious about these data. The EAC survey team was unable to get a reasonable level of survey response to the voting system malfunction questions, and the available data in the EAC survey report do not allow us to do any analysis as to whether the types of jurisdictions represented in these data presented in Figure 6.2 are at all representative of the broader population of election jurisdictions. It is certainly possible that election jurisdictions with substantial voting system malfunctions in the 2004 presidential election avoided answering this particular question in the EAC Election Day Survey, or those jurisdictions using certain types of voting systems tended not to retain or report these types of data.

The second effort to collect incident data arose in the spring of 2006, in a unique project conducted by the Election Science Institute (ESI) for Cuyahoga County (Ohio), a project in which we had the opportunity to be involved. Cuyahoga County in early 2006 was initiating the roll-out of its new Diebold touch screen voting system, used for the first time countywide in its May 2006 primaries. The ESI effort involved a number of different efforts to study the performance of the Diebold voting system used in Cuyahoga (which was equipped with a voter-verified paper audit trail), including polling of voters and election workers, an attempt to audit the voting machines used in the election, and for our purposes in this chapter, an extensive analysis of reports of incidents from poll workers involved in the election day touch screen roll-out.[9]

The Cuyahoga incident report data are unique and shed light on the many issues that can—and that did—arise in this primary election. Our research team (involving the two of us, and Caltech colleagues Jonathan Katz and D. Roderick Kiewiet) received incident report forms from 1,217 precincts (each precinct was provided an incident report form). From these completed reports, we built a database of 6,285 cases of incidents. We then coded each reported incident into a variety of incident categories and analyzed the incident report data by both incident type and precinct. In 14 percent of the incident reports, we found that there was either no incident actually reported (the report was blank) or a positive incident reported. Of the positive incidents, most were simply a statement that some task was performed by the election officials or that they precinct had been visited or called by the board of elections, vendor representatives, or technicians.

The distribution of reported negative incidents was sweeping: we identified four important substantive areas as showing the most reported incidents. First, about a third of all reported incidents (30.1 percent) involved the voter registration process. Many of these problems were associated with inaccuracies in the voter registration list used in the precinct. Second, about a fifth of problems (22.6 percent) were administrative in nature, including missing supplies, inadequate training, or a variety of procedural questions and problems. Third in significance were voting equipment problems, constituting 16.2 percent of reported incidents—which included problems with the voting machines, the printers, and the ballot access cards and encoders. The fourth, substantive area regarded election workers, who were the source of nearly one out of every ten reported incidents (9.5 percent).[10]

At first pass, even this level of analysis of the Cuyahoga primary incident report data leads to an important question about how significant the problems relating to the voting machines themselves were. On the one hand, more than 60 percent of the reported incidents arose because of issues that are not associated directly with the electronic voting devices used in the primary election, but resulted from problems with the voter registry, the administration of the election, or polling place workers. But on the other hand, it is clear that the roll-out of the new paper-trail equipped electronic voting devices was not problem-free: 16 percent of the reports did involve problems relating to the voting machines.

Detailed analysis of the voting machine problems resulted in the discovery that there were four different sources of problems with the machines: specific voting machine incidents, problems with seals, printer malfunctions, and problems with memory cards or their associated encoding devices. Of all the incidents reported, 8.4 percent dealt with some form of machine problem or failure. Nearly 40 percent of the machine problem reports were simply indications that one of the precinct's voting machines was not functioning, and nearly a quarter of other machine problem reports concerned the machine's memory cards. The problems with seals on voting system components were 4.2 percent of all reported incidents, and these were particularly problematic. If seals were missing or broken on the voting machine or its components, then the integrity of the ballots cast on that particular machine and in that particular precinct could be called into question, as seals are important methods of preventing and detecting tampering.

Problems with the machine printers were also a small, but distressing, fraction of reported incidents (4.2 percent of all reported incidents involved printers). Most (over 50 percent) of the reported printer problems were stated as printer failures, and a third of the reported printer problems were paper jams. Of course, the paper audit trail is the most important method of voter verification on the electronic voting systems used in Cuyahoga County, and because they constitute the ballot of record in Ohio, these printer problems, while clearly of a low incidence rate in this election, do provide data about the potential problems associated with the paper audit trail—problems that need further research.

Finally, of the machine-related incidents, there were problems reported that related to the voter access cards and the card encoder devices. The Diebold Accuvote-TSx, used by Cuyahoga County, requires that the voter obtain an access card (which looks much like a credit card or a hotel room access card); these access cards are formatted with the appropriate ballot and precinct information, and the voter puts the card into the voting machine in order to active at the machine for voting and to obtain the correct ballot style. In the incident reports, nearly 4 percent of all reported problems involved access cards or problems with the card encoder devices. Many of these reports were generic, and it is difficult to know how many voters were adversely affected by these problems, but about four of the ten reported access card or encoder problems involved general problems with access cards, and 20 percent concerned general problems with encoder devices. Almost another 20 percent of the reported incidents of this type stemmed from access cards that simply were dead and not functioning, while more than 15 percent related to nonfunctional card encoders.

The Cuyahoga County precinct incident reports from its May 2006 primary have yielded a wealth of useful data about the problems that the county faced in the first roll-out of its new electronic voting system, and no doubt there will be additional research of these data. Some of these problems appear to be inherent in the technology itself, including malfunctioning printers, access cards, and access card encoders, although from the incident report data as collected in Cuyahoga County it is impossible to know how many voters were affected by these problems on election day; and the incident reports do not contain information on how or whether these problems were resolved. Other machine-related problems relate to how voters or election officials used the devices, and some (Such as missing or broken machine seals) have procedural or administrative solutions. The Cuyahoga incident reports provide some of the only finely-grained data we have about how electronic voting systems fare in their initial use, and should help guide future policy making and research activities aimed at improving these voting technologies and their deployment.

OTHER PROBLEMS WITH E-VOTING IN 2004 AND 2006

Even within a voting technology class, different systems often vary in their performance. For example, data from Georgia (Bullock and Hood 2002) found that voters using an optical scan ballot who filled in an oval to record their vote total produced a lower residual vote rate than a voters who connected two arrows to complete their vote choice. Similarly, an ongoing study of voting technologies by scholars at the University of Maryland, University of Michigan, and University of Rochester has found that citizens have diverse experiences when they interact with different electronic voting technologies.[11]

When we consider critiques of various voting technologies in 2004, we break the problems with electronic voting into six distinct categories. We also note when similar problems occurred with other voting systems. No voting system performed flawlessly in 2004.

Problem 1: Adding and Losing Votes

There were two major electronic voting technology failures in the 2004 general election, both of which affected vote totals. One failure in Ohio gave 3,893 extra votes to Bush and a different failure in North Carolina lost 4,503 votes. In the first case, Franklin County, Ohio—which used the ELECTronic 1242 system, an older-style electronic voting system purchased in 1992 and manufactured by Danaher Control—produced unofficial results that had President George Bush receiving 4,258 votes compared to 260 votes for Democrat John Kerry, in a precinct where only 638 voters cast ballots. According to the director of the Franklin County Board of Elections, the problem occurred when the card from a voting machine was plugged into the counting software on a laptop, producing a flawed set of results.

In North Carolina, the problem again was with an electronic voting machine manufactured by a very small provider, UniLect. The problem was that the election officials had been told that a voting machine could hold 10,500 votes in storage, when the storage limit was actually 3,005 votes, given the ballot length. In early voting, a single machine was used by 7,535 voters; the storage unit successfully captured all votes up to its storage limit, but the 4,530 votes that were cast after the storage unit was filled were not captured. According to the manufacturer, the system was designed to flash a warning when the storage unit reached its capacity. However, the machine was not designed effectively to prevent this problem.

Problem 2: Poll Worker Training Problems

Orange County, California, consolidated some precincts—that is, put multiple precincts in a single polling place—in its 2004 primary elections. In a consolidated polling place, all of the precincts in the polling place may not have the same ballot style. For example, if precincts P1 and P2 are in a single polling place, P1 may have a ballot style that includes a race for school district A but P2 may have a ballot style that includes a race for school district B. If poll workers are not careful, they may give all voters the P1 ballot style, which means that the vote totals in the races for school district A will include votes from voters who are not eligible to cast votes in the race, and school district B will be missing votes from a given precinct.

In fact, approximately 1,500 voters were given the wrong ballots by poll workers who were using the county's new electronic voting system for the first time, and approximately 5,000 voters had their ballots tabulated in the wrong precinct. According to the *Los Angeles Times*, twenty-one precincts had acute problems, where more ballots were cast in the precinct than there were registered voters (Herndon and Pfeifer 2004). The impact of the error was broadly felt, with five U.S. congressional races, four state senate races, and five state assembly races affected by the error. The problem was clearly related to poll worker training, not the voting system. As the Associated Press (2004) reported, "Several poll workers said they didn't know more than one precinct had been assigned to their polling place, however, and thus gave some people the wrong access codes."

This case is illustrative of how voting technologies interact with election procedures. The problem was not the failure of the voting machines—in fact, the same problem could have occurred with any voting technology—but the performance of the poll workers. As a result of procedure failures, voters were not given the correct ballot for their precinct. When training poll workers, the issue of consolidation should have been a focus of the educational efforts.

Problem 3: Poor System Implementation

There have been many cases of poor implementation of new electronic voting systems since the 2000 elections, including one in Miami–Dade County, Florida. In two separate reports, the Miami-Dade County inspector general outlined in detail the problems that arose in this procurement and deployment. The first report (MDOIG 2002) concerned the Miami-Dade County primary election, the first election where the county's new electronic voting system was deployed. The second report (MDOIG 2003) examined the issues associated with the contract the county had with Elections Systems and Software (ES&S). One reason that Miami-Dade County wanted to move to touch screen voting is that it is covered by the language minority provisions of the Voting Rights Act. Under the law, the county has to provide voting materials, including ballots, to certain voters—those voters who speak Spanish or Creole—in their native languages.

The county did not pilot-test the new system before the primary election and encountered numerous problems in the system's implementation. Hundreds of polling sites did not open on time because of problems with the new equipment. The voting machines took roughly six and a half minutes to be initialized at the polling sites, with machines containing audio ballots requiring twenty-three minutes to boot. Additionally, every machine had to be booted in sequence; it was not possible to boot all machines in a precinct simultaneously. Because the original directions indicated that it would take only one minute to boot up the voting machines using a flash memory card, many poll workers removed the cards before the process was complete, causing it to fail (MDOIG 2002). As a result of the problems with the equipment, 23.6 percent of all poll sites did not open on time at 7:00 a.m. Four hours later, thirty-two poll sites—4.24 percent of all sites—were still not open, leaving more than 40,700 registered voters without a place to vote (Merzer, James, and Chardy 2002). Many poll sites that did open on time did so without the full compliment of equipment, creating long lines that led many voters to leave and not cast ballots.

In the aftermath of this election debacle, the Miami-Dade Inspector general found several severe flaws in the procurement process (MDOIG 2003); specifically, Miami-Dade County assumed that the state certification process for voting systems provided some guarantee of the quality of the system, an inaccurate assumption. As the MDOIG (2003, 8) notes in its report, the state

> sets forth the minimum standards of voting systems and establishes testing and certification procedures to determine if the required minimum standards have been met. . . . But what [the state] does not evaluate is the ease and operation of the voting system. . . . Certification is by no means a product endorsement; it has no bearing on whether it is a good product. . . . [or] whether the product will perform to the needs or expectations of any county. . . . Most importantly, certification does not test or verify the representations—sales pitch—made by the vendor.

Problem 4: Poor Election Auditing and Record Retention

Miami-Dade County, Florida was also the source of a farcical scandal regarding the auditing and retention of data from an election conducted using electronic voting technology. In July 2004 it was reported that the county had lost all of the electronic records from the 2002 gubernatorial primary—specifically, the event logs for each machine, which are a record of all activity on the machine from start up to shut down, and the ballot images from the election—as the result of a system crash. By law, data from the election are required to be retained for at least twenty-two months after the election (Goodnough 2004).

In response to the problem, the voting system manufacturer stated that the records retention problem was one of "state and local election management processes and procedures," not one of technology, something with which local election officials agreed. Had the county backed up the data correctly, they would not have been lost (Therolk 2004). Ironically, the data were not lost forever—but only "lost" on a bookshelf in a conference room next to the election director's office. Again, this is an example where election processes and procedures prove to be an ill-fit with a new voting technology. The county needed to have adopted and implemented specific procedures for handling and retaining electronic data in a safe form on multiple media concomitant with the adoption and implementation of the new voting technology. This clearly did not occur, resulting in what the *Washington Post* deemed to be a "surreal" story of lost and found data (Roig-Franzia 2004).

Problem 5: "Going Cheap" in System Deployment

The report by the Democratic National Committee regarding the 2004 election in Ohio illustrates the problem of "going cheap" in system deployment by not purchasing enough electronic voting machines to serve the voting population in a precinct. Specifically, in section VI of the report, Walter Mebane and Michael Herron find a relationship between the number of voting machines per person in a precinct and turnout in that precinct, with an increase of approximately 3.6 percent associated with a move from the first quartile of voting machines per registered voters to the third quartile.[12] By not having an adequate number of voting machines in each precinct, voters had to wait in longer lines than is appropriate, which deterred some voters from casting a ballot, reducing turnout. Mebane and Herron also find that the fewer machines in a precinct, the higher the residual vote rate in the precinct, as voters are deterred from reviewing their ballot carefully because of the pressure caused by the crowded precinct. By going cheap, not only do turnout rates decline but the quality of the votes cast also declines.

Problems Aplenty: The Great Debate about Anomalies

For those of us who are veterans of the voting technology debates, the days and weeks immediately following the 2004 presidential election were a heady—and quite frustrating—time. The flood of emails and telephone calls from the media, concerned citizens, and colleagues (and even the mothers of some of our colleagues!) seemed nonstop, and for weeks following the election it seemed as if there was essentially an "allegation of the day" circulating. Stoked by new technologies like the Internet and blogging, many of these allegations arose, peaked rapidly, but took a long time to fall off into the background of Internet chatter.

There were three main allegations of anomalies or irregularities associated with electronic voting in the 2004 presidential election. The first arose on election day and has been one of the most persistent of the allegations of irregularities—what we call the "Great Exit Poll Debate." On the afternoon of the election (West Coast time), we both began to receive phone calls on our mobile phones (as we both were "in the field" observing polling place practices on election day) informing us that preliminary releases of national exit poll data were pointing to a potentially surprising Kerry victory, with Kerry supposedly far ahead in these preliminary releases in some of the critical battleground states. These preliminary releases of the national exit poll data from the media consortium poll (conducted by Edison/Mitovsky Research for the major national networks and newspapers) were posted on a variety of Web sites. Adding to the confusion, the exit poll numbers that started coming out from the media outlets subscribing to the consortium polling data after all polls closed continued to show an apparent discrepancy between the exit poll results and the actual election outcome, with the exit poll results pointing to a stronger Kerry showing than the election results being tabulated that evening and in the days following the election.[13] Much of the analysis of these discrepancies has been produced by Freeman and Bleifuss (2006).

We are not going to attempt here to dive into the great debate itself but only to dig quickly into one aspect of it—whether these anomalies are associated with electronic voting machines.[14] On this particular question, there have been a series of analyses of the reported exit poll data, looking at whether there are discrepancies between exit poll results and tabulated voting results that are specific to or of a certain type in places using electronic voting equipment. One of the early studies in this area, relying on what the media had reported to be the state-by-state exit polls, was produced by the Caltech/MIT Voting Technology Project (2004a), and this report concluded that there was no evidence in those data to support the claim that there were clear correlations between discrepancies and voting systems used by states. A second report, produced by Edison/Mitofsky (2005), examined this question in more detail using data not available to researchers immediately following the election and found that "precincts with touch screen and optical scan voting have essentially the same error rates as those using punch card systems" (39). We have not seen additional analysis that indicates that the exit poll discrepancies were necessarily associated with electronic voting systems per se, as the available evidence continues to indicate that discrepancies exist across all types of precinct voting equipment used in the 2004 presidential election. Efforts to study the 2004 exit poll data should continue, and exit polling procedures need to be improved before the next presidential election.

The second major controversy regarding election anomalies and electronic voting systems popped up on the Internet right after the presidential election, and we call this one the "Great Florida Debate." This allegation, involving some analyses of differences between presidential partisan votes and voter registration statistics in Florida counties, stated that these discrepancies were the greatest in Florida counties employing electronic voting equipment. On November 10, 2004, the Caltech/MIT

Voting Technology Project (2004b) issued a report on this allegation, because this was a data phenomenon that political scientists involved in the VTP project were familiar with: many counties in the South with significant Democratic registration have been voting heavily Republican at the presidential level for some time. Other political scientists reached the same conclusions.[15] This controversy appears to have quickly died down following the election.

The third controversy arose in the middle of November, coming as a prominent social scientist at the University of California, Berkeley, and colleagues released a complicated statistical study of voting in Florida that alleged again that discrepancies in Florida appear to be associated with the use of electronic voting equipment.[16] This third controversy over alleged anomalies we call the "Great Berkeley Claim." Given the prominence of the research group releasing the study, a number of academic teams across the nation raced to examine the statistical analysis right before the Thanksgiving holidays. A number of scholars pointed out that the basic result, that discrepancies in favor of Bush existed in Florida counties using electronic voting machines in 2004, was dependent on "the peculiar and seemingly arbitrary functional form they use" (Brady et al. 2004, 11). Independent studies produced by three different academic teams all tended to refute the central Berkeley claim (see Sekhon forthcoming; McCullough and Plassman 2004; Wand 2004). The Great Berkeley Claim seems to have been put to rest in December 2004, though we periodically have heard reference to it on the Internet or discussions about election anomalies in the 2004 presidential election.

We could continue, as these great debates do continue, and in some circles there continues to be distrust and concern over the fairness of the 2004 election, with questions still being raised by some about whether electronic voting machines are the culprit "explaining" the election outcome. At this point in time, the evidence offered to date simply does not support these claims. While it might not convince our readers to hear this, we promise that we would be among the first to go public with any evidence regarding real anomalies, fraud, or indication of malfeasance regarding electronic voting machines and this election. We could make quite a career for ourselves were it the case that such evidence existed, and we suspect that had such evidence existed that both John Kerry's advisers and the Democratic Party more generally might not have thrown in the towel so quickly after the November 2004 election.

A Lack of Problems: Under the Magnifying Glass in California

In what might be a delicious irony for many—as California is a high-tech center of the world, with many of the major technology industries either headquartered or with operations in the state—California has in some ways been ahead of the nation in the debate about electronic voting. This prominence of California in the e-voting controversy is deeply rooted in the actions of two successive California secretaries of state (the state's chief election official). The first of these two secretaries, Bill Jones, was in office during the 2000 presidential election. While the problems that arose in Florida did not arise in a similar way in California, there was concern in some circles after the 2000 presidential election as many of the state's larger election jurisdictions used punch card voting systems, including card types that were both prescored and not prescored.

These concerns led a group of advocacy organizations (led by Common Cause and the American Civil Liberties Union) to file a lawsuit in January 2001 against Secretary Jones to force him to decertify the punch card voting systems; as secretary of state, Jones had it within his power to decertify any voting system, effectively meaning that it cannot be used any longer in California. As a result of the lawsuit, Secretary Jones decertified prescored punch card voting systems in California, in September 2001, and then both parties to the lawsuit began to debate when the decertification order should take effect, with Jones siding with the county election officials in seeking a deadline of July 2005, while the advocates looked for elimination of punch card voting systems in early 2004, before the state's spring primaries. In early 2002 the federal court judge agreed with the arguments of the advocates and gave the nine California counties using prescored punch card voting systems two years (early 2004) to replace those voting systems with newer voting technologies.

At the time, it appeared to many that it was inevitable that these nine affected counties would simply replace their prescored punch card voting systems with electronic voting machines, once state and federal money were available for punch card replacement. In fact, some of these affected counties, including the nation's largest election jurisdiction, Los Angeles County, had already begun studying electronic voting through pilot projects and by 2002 were well down a path toward making a massive transition from prescored punch card voting to touch screen voting systems.[17] As the critics of e-voting had now really begun to affect the elite debate about election reform in early 2002 (as we have discussed in earlier chapters), it appeared that California might implement electronic voting well before the rest of the nation.

All of this changed dramatically in 2002, once Kevin Shelley replaced Bill Jones as the secretary of state in California. Shelley, the son of a prominent San Francisco Democratic family (his father served as San Francisco mayor, in the state legislature, and in the U.S. Congress), was seen by political observers as a rising star in California Democratic politics; in his mid-thirties he was elected to the San Francisco Board of

Supervisors, moved quickly into the state assembly, and then, as a termed-out assemblyman, ran to replace Jones (who was himself a victim of term limits in 2002). Shelley quickly made a name for himself, in the state and nation, as he was tasked almost immediately after taking office with administering the novel and controversial recall election in 2003. Almost simultaneously, under pressure from both e-voting opponents and election officials across the state to provide some direction for the state's need to move to new voting technologies to replace prescored punch card voting systems, Shelley put in place an Ad Hoc Touch Screen Task Force. This task force produced a final report, in July 2003, which failed to satisfy voting rights advocates, county election officials, and e-voting opponents.[18]

The controversy continued to build in California, though the proximate issue of running the recall election occupied most of the attention of Shelley and the county election officials throughout the state—and what free time they had was put toward trying to get California's Help America Vote Act state planning process initiated (Alvarez and Hall 2005a). Almost immediately following the October 2003 recall election, Shelley surprisingly issued a series of directives in late November 2003 regarding electronic voting in California, the most important of which required that all electronic voting machines purchased in California after July 1, 2005, be equipped with a voter-verified paper audit trail, and that as of July 1, 2006, all electronic voting systems already purchased and in use be modified to allow for a voter-verified paper audit trail. Then, in February 2004, Shelley issued a series of additional security measures that he required to be in place in counties that were planning on using electronic voting devices in the 2004 primary and general elections.[19]

One of the primary security requirements was the use of "parallel monitoring," which Shelley's directive defined as "implementing a program to randomly select voting machines to be set aside for experts to vote on March 2, 2004. These machines will be voted exactly as if they were in polling places, any anomalies will be detected, and appropriate remedies will be pursued." The secretary of state did an identical parallel monitoring study in 2006 and again found no issues with the electronic voting equipment used in the state.[20] In effect, Shelley was putting in place a system like that used in other applications of electronic technologies—for example, in the gaming industry. In both Nevada and New Jersey, the government agencies in charge of regulating the gaming industries utilize parallel monitoring, in addition to a wide array of other security procedures, to ensure the integrity of electronic gaming devices.[21]

The parallel monitoring tests in California in 2004 provide another metric by which we can evaluate the performance of electronic voting systems in the most recent presidential election cycle. In the parallel testing done in the March 2004 primaries, the methodology involved testing a sample of each approved electronic voting system including electronic machines manufactured by Diebold, ES&S, Hart, and Sequoia. In the end, two electronic voting machines from a single randomly selected precinct in eight counties were selected for testing, and test scripts developed to mimic a variety of voting situations were utilized. The results of this first extensive parallel monitoring test in California revealed that the electronic voting systems randomly tested performed with perfect accuracy; all the discrepancies noted in the testing were attributed by the team to tester or script error or to provisional balloting variation—not voting machine error.[22]

The November 2004 parallel monitoring test was somewhat more extensive, involving again two electronic voting devices from a randomly selected precinct, but this time ten counties participated in the testing project. In this more extensive parallel monitoring test, in nine of ten counties the electronic voting systems performed again with perfect accuracy, and the discrepancies noted in those nine counties were attributed to testing problems, not to the voting systems themselves. One anomaly did appear in the initial testing in Merced County; there "the tester appeared to correctly 'tap' the screen to select candidate 'Bush' for President however, on the screen candidate 'Peroutka' was highlighted and the ballot recorded for him."[23] This one anomaly arose with an ES&S touch screen voting device; the device was apparently sealed until further analysis could be done, of both the voting machine itself and the videotape recording of the testing of the specific machine. The subsequent analysis of the voting machine indicated that testing error was responsible for this problem, not the electronic voting machine itself.[24]

Thus, the California parallel testing process from the 2004 presidential election cycle gives us important data that indicate a lack of problems, from this particular approach, in the electronic voting systems used in California. This testing involved a total of thirty-six (sixteen from the primary and twenty from the general election) electronic voting systems, representing the full array of electronic voting platforms certified for use in California. Each of these electronic voting systems had 101 test voting scripts run on them, resulting in a potential electorate of test votes that is relatively sizable (3,636). Of these 3,636 test votes run in the parallel test, the only problems noted were mistakes made by the testers or in the testing process—not a single problem in the electronic voting systems themselves was identified. Of course, there are two problems relying on data like these to conclude that the larger population of electronic voting machines used in the 2004 elections in California were perfectly accurate. One is that only a small sampling of electronic voting machines was tested; because we do not know the possible distribution of problems or security breaches more generally, we cannot really know at this point what the statistical power of this parallel test may be. The other problem is that we do not know the comparable rate for other voting technologies used in California, and so we cannot compare the performance of electronic voting systems with other voting systems— for example, optical scanning voting devices or the InkaVote system used exclusively in Los Angeles County.

The Final Problem: Glitches in 2006

The 2006 election illustrated a range of "glitches" that can occur in elections. Consider the following three examples, one from a paper-based voting system and two from electronic systems. First, in Idaho, the voters in Bannock County were given the wrong type of pens to mark optical scan paper ballots. As a result, the tabulation machines could not read the ballots. After the election, poll workers had to "remake" all of the ballots, marking voter choices so that they could be read by the tabulation machines. Second, in Utah County, Utah, more than 100 precincts were unable to program voter access cards using the encoders that they were given. Until poll workers converted one voting machine in each precinct to operate as an encoder, the voting machines could not be used. For approximately one hour from when the polls opened, the voters were unable to cast ballots on the machines.

The third problem arose in Sarasota County, Florida. There, the U.S. House race in the Thirteenth Congressional District to succeed Katherine Harris had an unusually high residual vote rate that was almost ten times the rate for the governor's race.[25] Additionally, as we discussed in an earlier chapter, the undervote rate was high only in Sarasota County (not in other counties that also are in the district), and only for voters who used Sarasota County's ES&S electronic voting system for early and election day voting (not for absentee ballots cast by mail). Although the simplest explanation for the problem is that the race was poorly laid out on the touch screen machine ballot, and was different than other races on the ballot because there were only two candidates listed, there are still questions as to why this race had high residual votes compared to others.

SUMMARY

This chapter presents an array of information about the performance of various voting technologies. We start by noting the problems that exist in the baseline technology: votes cast on a paper ballot. In the four years since the 2000 elections, the problems that have existed with paper ballots persist. Paper ballots are still lost and found, and counts change from the first count done by election officials to the second and third and fourth counts. This was clear in the 2004 governor's race in Washington State, where vote tabulations on electronic machines remained consistent over time but the paper ballot count kept changing as ballots were "found" and "voter intentions" were discovered.[26] Some critics argued that the problem with the electronic machines was that there was nothing to recount; the ballot counts they produce and the ballot images are exact. However, the continual counting and recounting and recounting of paper ballots, with new numbers produced with ballots lost and found, can also undermine confidence in the effectiveness of paper as a voting technology.

When we consider the residual vote data, we can see that switching from a historically poor performing voting technology to a historically effective voting technology is one way of improving the number of ballots that get counted. Optical scan systems perform particularly well in this analysis. Additionally, though, we see that, even within the same voting technology, something else occurred between 2000 and 2004. Three plausible explanations for the reduction in residual votes even in localities that kept the same voting technology are that voters learned to be more cognizant of potential problems with their ballots between 2000 and 2004, poll workers did a better job of operating polling places to ensure that errors were minimized, and election officials did a better job of educating both voters and poll workers about the possible problems.

Perhaps the most interesting issue with electronic voting is that the problems that often arise from its use have little to do with the machines themselves. The problems are often administrative. In the worst case, the election officials have no idea what they are really buying and fail to prepare adequately for the new election challenges posed by transitioning to a new voting system. However, the problems with electronic voting are typically more pedestrian. Election officials fail to train their poll workers regarding how to best implement these systems, to consider that they need to have the different internal capacities in order to retain electronic data, or to have the internal staff capacities needed to address the different issues that electronic voting brings to the administrative process.

The transition to electronic voting that occurred in many jurisdictions between 2000 and 2004 resulted in a marked reduction in residual vote rates in the communities that switched from any voting system to electronic voting technologies. These reductions resulted in thousands of votes being counted in 2004 that were not counted in 2000. Analysis of the transition to electronic voting in Georgia suggests that the votes of African Americans and low-income individuals are more likely to go uncounted on nonelectronic voting technologies; the transition to electronic voting reduced the residual vote rates markedly for both groups. Furthermore, exit poll data from the 2000 election indicate that both these groups were strongly inclined to support Gore and not Bush.[27] This raises an obvious question: what would have been the result if the voting in Florida in 2000 had been conducted on electronic voting technologies? Would a different president have been elected?

But the analysis in this chapter also raises some important methodological questions for social scientists and policy makers. We would all like to know, in each election, how voting technologies perform. But as our analysis in this chapter shows, performance data are not easy to obtain. There is now an established and productive research literature that focuses on the residual vote measure of voting system performance, and we rely upon that measure heavily in this chapter. But when it comes to other issues, for example, usability, security, integrity, and other aspects of reliability, comprehensive data are still difficult to obtain.

In the absence of comprehensive data on problems with voting systems, researchers typically rely heavily upon media reports and litigation, to shine the light on problem spots and to help provide data and research that we can all draw upon for our evaluation of voting technologies. These data, while useful, are not systematic and sometimes are not necessarily collected, analyzed, and presented in ways that can be used in academic work. Data like those the EAC attempted to collect, at the national level, in the 2004 election, have great possibility—though clearly the EAC's experience with collection of these data should provide some important lessons about how to do it better in the future. Other data collection exercises, like the Cuyahoga project in 2006, should be replicated in the same jurisdictions, and then repeated in other locations. We should move to a level of scientific study in this new and multidisciplinary field of studying voting technology, where we have access to systematic data on voting system performance, on a number of dimensions, across jurisdictions so that more in-depth datasets can be made available for a subset of jurisdictions.

APPENDIX: MULIVARIATE ANALYSIS OF SWITCHING VOTING SYSTEMS

In this analysis, we conduct what social scientists refer to as a multivariate analysis, where we examine a dependent variable—in this case, the residual vote rate—and determine what factors, such as switching equipment or switching to specific types of equipment, affect this dependent variable. Table 6.3 presents the results of four regression analyses, each one of these regression models employing a different specification so that we can see how robust our core results are across different

TABLE 6.3
Multivariate Analysis, 2000–2004 Presidential Elections

Independent Variable	Log(F(Residual Vote Rate))		Residual Vote Rate	
	1	2	3	4
Punch cards	***	***	***	***
Paper ballots	−0.461	−0.462	−0.010	−0.009
	(0.039)	(0.178)	(0.001)	(0.006)
Lever machines	−0.43	−0.586	−0.009	−0.009
	(0.039)	(0.111)	(0.001)	(0.004)
Optical scan	−0.339	−0.411	−0.007	−0.010
	(0.026)	(0.070)	(0.001)	(0.002)
Electronic	−0.245	−0.338	−0.005	−0.007
	(0.035)	(0.087)	(0.001)	(0.003)
Shift in voting technology.	−0.058	−0.128	−0.000	−0.003
	(0.024)	(0.045)	(0.001)	(0.002)
Governor or Senator on ballot	0.067	0.080	0.001	0.001
	(0.022)	(0.022)	(0.001)	(0.001)
Log(turnout)	−0.055	−0.052	−0.001	0.001
	(0.008)	(0.087)	(0.000)	(0.003)
Percent white	−0.005		−0.000	
	(0.001)		(0.000)	
Percent Hispanic	0.003		−0.000	
	(0.001)		(0.000)	
Age 18–24	−0.014		−0.000	
	(0.003)		(0.000)	
Age 65 and older	0.004		0.000	
	(0.003)		(0.000)	
Median income in 10,000s	−0.226		−0.009	
	(0.049)		(0.002)	
(median income)^2 in 10,000s	0.011		0.001	
	(0.005)		(0.000)	
Constant	−1.922	−2.996	0.082	0.020
	(0.136)	(0.816)	(0.005)	(0.027)
Observations	3825	3825	3825	3825
R^2	0.46	0.83	0.28	0.82
Fixed effects (not shown)	State and year	County and year	State and year	County and year

(Continued)

TABLE 6.3 (*Continued*)

Independent Variable	Log(F(Residual Vote Rate))		Residual Vote Rate	
	1	2	3	4
Number of categories	42	2,472	42	2,472
F test	F(41, 3769)	F(2471, 1345)	F(41, 3769)	F(2471, 1345)
	= 44.77	= 2.215	= 17.57	= 2.24
	(p = 0001)	(p = 0001)	(p = 0001)	(p = 0001)

model specifications. Equations 1 and 3 examine changes in residual vote rates from 2000 to 2004, including fixed effects for year and state, with punch cards used as the baseline or comparison case in each model specification.

The difference between these two statistical models is that in the first equation the dependent variable is a log of the residual vote rate, while in the third equation we use the residual vote rate itself without taking the log transformation.[28] Equations 2 and 4 also examine changes in residual vote rates from 2000 to 2004 but with fixed effects controls for year and county; regression model 2 uses the log transformation, while regression model 4 does not. Models 2 and 4, with county and year fixed effects, do not include the county-level demographic control variables in those analyses, as it is impossible to include those county-level variables while also controlling for county fixed effects. The results across all four specifications help confirm the basic, core results from our earlier and less technical analysis.[29] Each voting technology improves residual vote rates when compared to punch cards, as can be seen in the negative sign for each variable for machine type. We also see that switching systems resulted in a reduction in the residual vote rate in each of the four equations.

Chapter 7

PUBLIC ACCEPTANCE OF ELECTRONIC VOTING

In an announcement in the summer of 2004, Gilead Sciences—a leading biopharmaceutical company that develops and sells innovative antiviral drugs including Tamiflu—was added to the Standard and Poor's 500 Index, a widely watched stock market index. This stock index is designed to cover market-leading companies in leading industries, especially of the so-called large-capitalization component of the stock market.[1] Selection to be included in the S&P 500 index indicates recognition that Gilead is a top company in its industrial sector, a company that in 2004 generated total revenues of just over a billion dollars. In 2004 Gilead reported spending over $300 million in "selling, general and administrative" expenses (which rose to nearly $380 million in 2006).[2] It is difficult to know how much of this might have been spent on marketing and market research activities, but industry watchdogs have argued that marketing costs may be a large fraction of the overall expenditures of firms in the pharmaceutical industry.[3]

The importance of marketing to any corporation is such a given that today companies are now examining how their products fit into the life-styles of their customers with the goal of serving the life needs of customers in order to build loyalty (Seybold 2001). Behind this biopharmaceutical business, one that generates billions of dollars in revenues, there are substantial amounts of research done to determine how companies like Gilead should market its products, to identify who the consumers are, and to find how to best present its products to these consumers. For any company of Gilead's size and clear success so far, developing a clear marketing strategy for customers is a critical component of any business model.

We used Gilead as an example, because estimates are that the United States spent approximately a billion dollars annually on election administration in 2000, and probably even more than that today.[4] Although election administrators expend roughly the same amount in public revenues as Gilead receives in revenue, it is not clear that election administrators do anything like the market studies that corporations like Gilead undertake. In fact, we are not aware of many election administrators who think of running an election in terms of business models or marketing strategies and see voters and the public as the ultimate consumers of their product.

As a bottom-line issue, the views of the electorate and citizenry as consumers of the product of election administration are critical to understand. After all, if the citizenry does not believe in the fairness, accuracy, openness, and basic integrity of the election process, the very basis of any democratic society might be threatened. And in the wake of two disputed presidential elections in the United States, some assert that Americans are concerned about the state of their electoral process—that the citizenry may not trust the integrity of elections, especially with the increased use of electronic voting technologies.

For example, Richard Hasen, a top academic election lawyer, testified before the Commission on Federal Election Reform: "It should go without saying that public faith in the integrity of the election system is a cornerstone of democratic government. Yet the data are worrisome. According to a post-election NBC News/Wall Street Journal poll, more than a quarter of Americas worried the vote count for president in 2004 was unfair."[5] Congressman Rush Holt went so far as to title his legislation that would amend HAVA to require "voter-verified paper audit trails" the "Voter Confidence and Increased Accessibility Act of 2005" (HR 550), in effect implying that voters lack confidence in existing voting technologies.[6] In work that we have done with Morgan Llewellyn, we have studied voter confidence in vote counting, in surveys running from 2004 through 2006. In our most recent survey (fielded in late October 2006), we found that 22 percent of likely voters lacked confidence that their ballot was tabulated as they intended in the 2004 presidential election, which had risen from 11 percent in March 2005. We also documented in our pre-2006 survey that African American likely voters were less confident than whites and that Democratic likely voters were less confident than Republicans.[7]

Although we and other scholars like Hasen have mined public opinion polling data on election reform, there is a surprising dearth of detailed analysis of American opinions on the electoral process, voting technologies, election reform, and confidence in the integrity of the process.[8] The question of public acceptance of electronic voting, and the perceptions of the public about electronic voting more generally, remains largely uncharted territory. Although there have been scattered reports of polling data on some aspects of electronic voting, especially Internet voting, and some attention in media polls after the 2000 and 2004 presidential elections about public assessments of the fairness of the process, there are virtually no academic studies that we know of that systematically study public perceptions of electronic voting and election reform.[9] In this chapter, we report on public opinion surveys we have conducted from before the 2004 presidential election to just before the 2006 midterm election that asked respondents a series of questions about their perceptions of the electoral process and, more precisely, their opinions about electronic voting.

SURVEYING AMERICANS ABOUT ELECTRONIC VOTING

Our surveys about election technology and reform were conducted in August 25–29, 2004 (Wave 1), March 9–15, 2005 (Wave 2), January 18–24, 2006 (Wave 3), and October 26–31, 2006 (Wave 4). All four surveys were implemented, and the interviewing conducted, by the professional staff of International Communications Research (ICR), a well-established and well-known market and opinion research firm headquartered in Media, Pennsylvania. We took advantage of their omnibus survey methodology (EXCEL), which allows researchers to place a small number of specialized questions on a national probability survey, and to receive the full set of responses to their questions as well as a standard set of demographic data about each respondent.[10] Our Wave 1 survey contained the responses from 1,014 interviews; Wave 2 survey was the result of two successive EXCEL omnibus surveys, with 2,032 total interviews; Wave 3 was also the result of two omnibus surveys, with 2,025 completed interviews; Wave 4, immediately before the 2006 midterm elections, has the responses of 1,084 American adults, including an oversampling of 156 non-Hispanic African Americans. Here we study the responses of only those who said they were registered to vote and who said they have voted in 2004 (or were likely to vote in 2006 in our fourth survey wave).

Our research on the perceptions and evaluations of Americans about voting technologies is divided in this chapter into three sections. In the first, we consider survey data on the evaluations of Americans regarding electronic voting itself. Then, we turn to the question of how Americans evaluate auditability and voter-verified paper audit trails for electronic voting technologies. We conclude with a focus on the more general question of the confidence of Americans in different voting technologies.

EVALUATIONS OF ELECTRONIC VOTING

Prior to the 2004 election, there was little attention paid to the issue of public perceptions of electronic voting; what little polling was done following the 2000 election about voting technologies was usually focused on the longer-term question of Internet voting. For example, in a national probability telephone survey conducted in the spring of 2001, 31 percent of the respondents supported Internet voting, when asked "Do you favor or oppose a system that allowed Americans to vote in elections over the Internet"; 59 percent were opposed, and 10 percent had no opinion.[11]

As the 2004 presidential election approached, two survey studies were released that dealt with public opinion about electronic voting. One, released in March 2004 by M. Glenn Newkirk of InfoSENTRY Services, Inc., focused on a very straightforward question. Survey respondents (1,026 interviews conducted in early February 2004) were asked to evaluate how much trust they place on a variety of voting systems "to produce confidential and accurate election results." One of the voting systems described to survey respondents was "an all-electronic, computerized voting system that is commonly known among elections practitioners through the shorthand terms of Direct Record Electronic (DRE) and 'touch screen' systems."[12] On a 5 point scale, the mean score across survey respondents was 3.8 for electronic voting systems (40 percent of the sample, the modal response for electronic voting, said they had "very high trust"), 3.6 for precinct optical scan, 2.8 for vote-by-mail optical scan, and 2.7 for the Internet.

The second set of survey results was released in early September, by www.findlaw.com.[13] When asked about concerns with the accuracy and potential for fraud associated with electronic voting machines, 38 percent of the 1,000 survey respondents were worried about their accuracy, 61 percent were not worried about their accuracy, and 1 percent had no opinion or did not answer the question. As to the potential for tampering with electronic voting machines, 42 percent were worried, 57 percent not worried, and 1 percent had no opinion or did not respond. Even though the results of this survey indicates that clear majorities in the early fall of 2004 were not worried about either the accuracy or the potential for tampering with electronic voting machines, it is interesting to note that the press release put out by www.findlaw.com was titled "Many Americans Distrustful of Electronic Voting Machines, Says New FindLaw Survey." Our review of these data leads us to think that this press release would have been better titled "*Most* Americans *Trustful* of Election Voting Machines, Says New FindLaw Survey"—a title consistent with the basic finding in this particular survey.

Our approach to studying basic evaluations of electronic voting was to probe American opinions more deeply about the four major arguments made by both sides on the debate on electronic voting. As we have noted in earlier chapters, critics of e-voting assert that it makes fraud and unintentional errors more likely. Supporters of e-voting claim that it is more accurate and more accessible, especially for voters with disabilities or those who would rather vote in languages other than English. Accordingly, we devised four survey questions that measure evaluations of survey respondents for each component of these arguments, pro and con. We

posed these same four questions to all respondents in each of our four
surveys and rotated the order in which these questions were posed to
respondents to insure that the survey results were not biased by the order
in which the arguments were presented.

Our survey questions on electronic voting began with the following
introduction: "You may have heard discussion about the use of electronic
touch screen or direct recording electronic voting machines in the presidential election this fall. I'm going to read you some statements about
electronic voting, and want to know whether you agree or disagree with
each statement, or if you have no opinion." Respondents were then asked
to evaluate these four arguments:

- Electronic voting systems increase the potential for fraud.
- Electronic voting systems are more accurate.
- Electronic voting systems make voting easier for people with disabilities.
- Electronic voting systems are prone to unintentional failures.

We asked these questions in all of our four survey waves, and their
responses to these questions for voters are given in table 7.1. In the
responses, we find a quite nuanced portrait of how adult American voters
view electronic voting machines. On each dimension that we evaluated,
the modal response tends to be agreement—that is, more voters agreed
that electronic voting machines were more prone to fraud and more
prone to failure, but also agreed that electronic voting machines are more
accessible and more accurate. There is especially high agreement that
e-voting technologies are more accessible: across all four surveys, approximately 60 percent of voters agree that these technologies make voting
easier for people with disabilities. In each survey, fewer than 15 percent
of voters disagree that electronic voting technologies help make voting
easier for the disabled.

For the other three dimensions, more voters agree than disagree with
each statement, though the relative difference between agreement and disagreement is not as large as on the accessibility dimension. Across all four
waves, more voters agree than disagree that electronic voting systems are
more accurate; but more voters also agree than disagree that electronic
voting systems are prone to unintentional failures. But when it comes
to whether electronic voting systems increase the potential for fraud,
opinions of voters are mixed; in our first wave (August 2004) 36 percent
agreed with that statement, and 29 percent disagreed—which is very similar to opinions in our October 2006 wave, where 38 percent agreed that
electronic voting systems increase the potential for fraud, and 29 percent
disagreed.

Furthermore, and a point that has not been adequately noted in previous discussions of American evaluations of e-voting, the public exhibits

TABLE 7.1
Opinions of Voters about Electronic Voting, 2004–2006, %

	Voters			Agree-Disagree
	Agree	Disagree	No Opinion	
Electronic voting systems increase the potential for fraud				
August 2004	36.4	29.1	34.5	7.3
March 2005	41.0	31.5	27.6	9.5
January 2006	36.5	34.3	29.1	2.2
October 2006	38.0	29.1	32.9	8.9
Electronic voting systems are more accurate				
August 2004	41.1	25.8	33.1	15.3
March 2005	39.7	26.2	34.1	13.5
January 2006	45.6	23.9	30.5	21.7
October 2006	43.9	21.8	34.3	22.1
Electronic voting systems make voting easier for people with disabilities				
August 2004	62.0	11.2	26.7	50.8
March 2005	60.1	14.4	25.5	45.7
January 2006	61.8	12.3	25.9	49.5
October 2006	57.3	14.6	27.1	42.7
Electronic voting systems are prone to unintentional failures				
August 2004	43.2	23.4	33.3	19.8
March 2005	45.5	21.6	32.9	23.9
January 2006	43.0	24.8	32.2	18.2
October 2006	44.2	23.4	32.4	20.8

a large amount of uncertainty—or possibly ambivalence—about each evaluative dimension. Our survey questions gave respondents the option to say that they did not have an opinion, and between one-quarter and one-third of respondents in each case said they had no opinion. This no-opinion response could reflect a lack of exposure to debates about e-voting, a lack of exposure to this voting technology, or a simple lack of interest in the electronic voting debate.

As our survey waves span slightly more than two years, during which there was substantial discussion in the media and elsewhere over the merits of electronic voting, we have a unique opportunity to gauge whether that debate has had much influence on the opinions of voters about electronic voting. In many major newspapers, in many Internet "blogs," and among the political-oriented chattering classes, electronic voting was a contentious debate at this time, a din that we discussed earlier in this book. The interesting question here is the extent to which this elite debate about the merits of electronic voting, both before and after the 2004 presidential election, penetrated into the perceptions and opinions of American voters.

Unfortunately, we do not have what survey methodologists call "panel" data—survey responses from the same individuals, collected at different points in time. Such data would be ideal for testing whether the 2004 election, and the e-voting controversy that swirled around in elite debates right before and after that election, had any impact on the attitudes of Americans. Instead, we have "cross-sectional" data—data with independent samples, but identical questions, posed at different points in time. While less ideal, the data allow us to test whether the overall distributions of e-voting evaluations have shifted over this time period, potentially as the result of the elite debate about electronic voting.

Table 7.1 shows the responses of voters for each survey wave and in the final column computes the difference between agreement and disagreement. If we assume that the elite e-voting debate did penetrate the public, there are two different hypotheses we might test about changes in public opinion of e-voting. The first is that, to the extent that the debate about e-voting was received and digested by the broader American populace, we should see a reduction in uncertainty about e-voting's attributes.[14] A second hypothesis about what we might observe in table 7.1 would be a shift in evaluations away from a positive evaluation toward more negative evaluation. That is, our basic hypothesis is that we should observe a larger fraction of survey respondents in the final survey wave (October 2006) than in the first wave (August 2004) who see e-voting as fraud- and error-prone, and fewer who see e-voting as more accurate and more accessible, if Americans are following the lead of media coverage of the e-voting debate as we depicted it earlier in this book.

Contrary to both hypotheses, we see no strong changes in the general tenor of voter assessments about electronic voting. First, and perhaps most telling, we do not see much reduction in the uncertainty or ambivalence of voters regarding their assessments of electronic voting on any particular dimension. In each case, we see that the rate at which voters said they had no opinion regarding electronic voting's accessibility, accuracy, and potential for fraud and unintentional failure is essentially unchanged. This indicates that overall it is not clear whether the debates over electronic voting that raged in the media and on the Internet between 2004 and 2006 served to necessarily reduce the amount of uncertainty or ambivalence that many American voters had regarding these new voting technologies.

Second, for two of the evaluative dimensions that have received the most media attention—whether electronic voting systems increase the potential for fraud, and whether they are prone to unintentional failures—we see little change between 2004 and 2006 in the difference between agreement and disagreement. In the first case, regarding the potential for fraud, in August 2004 36 percent of voters agreed that e-voting increased the potential for fraud and 29 percent disagreed, a difference of 7 percent. In October 2006, 38 percent of voters agreed with this statement and 29 percent of voters disagreed. This difference of 9 percent is a slight change over the August 2004 result, but the percentages agreeing and disagreeing are well within the survey's sampling error of 3 percent. The same relative stability exists for voter opinion about whether electronic voting systems are prone to unintentional failure, with about 20 percent more voters agreeing with that statement than disagreeing in each of our four survey waves.

In two of the dimensions however, we do see evidence of changes across this two-year period. First, we see a slight decrease in sentiment that electronic voting systems are more accessible, and a slight increase in disagreement with that statement. The agreement-disagreement difference for the accessibility question was nearly 51 percent in the fall of 2004; the difference is 43 percent in October 2006. There is also a slight change in the agreement-disagreement different for the accuracy of electronic voting systems. In August 2004, 41 percent agreed that electronic voting systems are more accurate and 26 percent disagreed, which is a 15 percent difference. By October 2006, 44 percent of voters agreed and 22 percent disagreed, producing an agree-disagree difference then of 22 percent. Although the change in those agreeing is only 3 percent, which is slight given the survey's margin of error, the decline in those disagreeing is 4 percent and the overall change in the agreement-disagreement value is 7 percent.

Our survey data indicate that, in many ways, the American electorate may be undecided about the merits of electronic voting. At this point, the debates that have raged in the media and on the Internet do not seem to have had a substantial effect on the opinions of American voters. We continue to see high rates of uncertainty and ambivalence regarding the relative merits of electronic voting, and this high degree of uncertainty and ambivalence has not changed appreciably since 2004. We see little consistent change in the electorate's view of whether electronic voting increases the potential for fraud, or whether these voting technologies are more prone to unintentional failures. We see some decrease in agreement with the notion that electronic voting systems are more accessible for the disabled, but some increase in the idea that electronic voting systems are more accurate. In sum, our data present a portrait of an electorate in which many voters are still evaluating these new voting technologies.

VIEWS ON PAPER AUDIT TRAILS

Concerns about fraud, errors, accuracy, and accessibility may loom large in many of the elite debates about electronic voting. However, our survey data show clearly that there is no unified "public opinion" about the basic merits of electronic voting. Despite this lack of public clarity, one segment of the advocate community and some prominent voices in the media have called for a very specific solution to these problems. This solution, which is rapidly becoming law in many states and is now being implemented in jurisdictions throughout the nation, is the so-called voter-verified paper audit trail (VVPAT) that we discussed earlier in this book.

There have been two early public opinion studies of VVPAT technologies. Both were conducted in Nevada, which was the first state to implement a VVPAT system. The first survey was conducted by the University of Nevada, Las Vegas, in 2004. It found that 85 percent of voters support the concept of voter verification on electronic voting machines, but also found that one in three voters did not use the verification process themselves.[15] Moreover, the survey also found that voter verification raised concerns about the ability of voters to read the print through the small screen that covers the paper, as well as concerns regarding how such systems could increase lines. Voter verification was estimated to add one to five minutes to the time it took a voter to cast a ballot, which could prove substantial in a high-turnout election. A second survey was conducted using exit polling in the 2004 general election in Nevada by a private research firm that was contracted by VoteHere.[16] This study found that 31 percent of voters either did not know what the voter verified paper trail was on the voting system they had just used or had not even noticed the paper system at all. Of those who did notice the paper trail, 14 percent did not use the paper trail to verify the ballot. Of those who did use the verification system, only 52 percent actually checked all of the races on the ballot, and 38 percent checked some of the ballot. One of the most interesting findings in this exit survey is that, given the choice between engaging in voter verification in the polling place or engaging in voter verification using a paper receipt—like those provided by cryptographic verification systems—the voters chose the latter by 60 percent to 36 percent.

Given the prominence of the VVPAT solution in elite debate, we thought that we should see how well-formed the opinions of American citizens are about the desirability of VVPAT, and elicit their opinions on what form of paper-based audit trail they might desire if they are to use electronic voting devices. Accordingly, in the second wave of our survey project we developed and implemented three questions about VVPAT; these questions were posed to a randomly selection subsample of half of

TABLE 7.2
Voter Opinions about VVPAT, %

	Voters: Opinions about VVPAT
If you had a choice, would you rather vote using a paper ballot or an electronic ballot?	
Paper ballot	51.3
Electronic ballot	46.3
Do not know	2.2
Refused	0.3
If you had a choice, would you rather vote using an electronic ballot that has a paper audit trail or an electronic ballot without a paper audit trail?	
Electronic ballot with paper audit trail	81.2
Electronic ballot without paper audit trail	11.2
Do not know	7.3
Refused	0.3
If you were to vote using an electronic ballot, which would you most prefer?	
No paper audit trail	12.8
A paper receipt that you can take home	41.7
A paper receipt you leave	41.5
Do not know	3.7
Refused	0.3

our second-wave survey respondents.[17] The three questions on VVPAT that we developed and implemented were:

- If you had a choice, would you rather vote using a paper ballot or an electronic ballot?
- If you had a choice, would you rather vote using an electronic ballot that has a paper audit trail or an electronic ballot without a paper audit trail?
- If you were to vote using an electronic ballot, which would you prefer the most?

We present in table 7.2 the basic responses to these three questions by voters in our second-wave sample.

Beginning with the simple issue of whether American voters would prefer a paper or an electronic ballot (without any other detailed options provided), the electorate is evenly divided. Fifty-one percent would prefer a paper ballot, but 46 percent would prefer an electronic ballot, and only around 3 percent did not state an opinion or did not answer this question.

Thus, American voters appear to be evenly split on the basic question of whether they prefer paper or electrons for their voting medium, as stated in these somewhat abstract terms.

When probed further about electronic voting and when asked expressly whether they want to vote electronically with or without a paper audit trail, more than 81 percent of voters in our survey agree that they would like an electronic ballot with a paper trail. Only 11 percent would prefer electronic voting without a paper audit trail, with about 7 percent having no opinion or not answering the question. This underscores the support for paper audit trails by American voters, when directly asked about them in the context of electronic voting. This sentiment is probably part of the reason that the VVPAT has seen so much debate by state legislatures in many of the states, and why as of 2006 twenty-two states required that voting systems produce a VVPAT.[18]

Our last question in this set focused on how voters might want the paper audit trail: a paper receipt that must be left in the precinct, a paper receipt they could take home, or no paper audit trail at all. Although nearly 13 percent of voters stated they preferred to have no paper audit trail at all, we see in table 7.2 that roughly 42 percent of voters wanted to leave their paper receipt in the precinct—the implementation of VVPAT that appears to be most favored by its advocates and by election officials. However, an identical percentage of voters (almost 42 percent) stated that they wanted a paper receipt that they could take home with them. This implementation of VVPAT for electronic voting requires using cryptographic, electronic verification systems, such as those that have been developed by corporations like VoteHere, Democracy Systems, and Scytl. These systems have not been advanced anywhere in the United States, even though such systems are thought by some to produce better verification and have been pilot tested in European elections.

Our new data on the preferences of Americans about VVPAT are quite illuminating and are in many ways at apparent odds with the opinions of the VVPAT advocates and election officials. It appears that, as a general matter, Americans are evenly split about whether they want to vote electronically or on paper. However, when given the opportunity to say whether they would want to vote electronically with or with out a paper audit trail, most Americans prefer electronic voting with the paper audit trail. Yet, by a near majority, Americans want to take their paper receipt home with them, most likely as some lasting proof that they voted. This latter sentiment is one that we have also verified in our qualitative observations of voters who are using touch screen voting systems equipped with VVPATs: a surprisingly large number of these voters either try to access the VVPAT to take home or ask polling place workers why they can not have it.[19] This suggests that election officials and voting system vendors might work to better educate voters about the VVPAT, or to investigate the production of some sort of ballot "receipt" that voters can take with them after they vote, though in the latter case it is unclear at this point whether that "receipt" should have any information on it regarding how voters cast their ballot.

A DIVERSITY OF OPINIONS ABOUT E-VOTING

Like most matters in American politics and society, there is considerable diversity across the public in evaluations of electronic voting and the VVPAT. Because there is almost no extant research before ours to rely upon to understand how different groups of Americans react to these questions, our analysis of how evaluations of electronic voting and VVPAT vary across the American public is more inductive than deductive, more like "data mining" than systematic, theoretically driven public opinion research. We present the evaluations of electronic voting of the voting public across the four dimensions in our four surveys wave in table 7.3 and 7.4, but for only two subgroups of the electorate: we examine e-voting evaluations by voter partisanship and race. Partisanship and race have loomed large as factors that seem to produce different evaluations of voter confidence. We are interested in evaluating whether these same voter attributes are producing differences in evaluations of electronic voting.

Table 7.3 gives the percentages of voters who identify as Republican, Democrat, or Independent, who agree with each evaluative statement. The final two columns of the table give the differences in the percentages of Republicans and Democrats who agree and the percentages of Republicans and Independents who agree.

As we see in table 7.3, there are substantial partisan differences in the evaluation of electronic voting technologies. Beginning with the lightning rod issue—whether electronic voting systems increase the potential for fraud—we see that a growing percentage of Democrats agree with that perception. The Republican-Democratic difference has grown from −3.0 in August 2004 (3 percent more Democrats than Republicans thought electronic voting increased the potential for fraud) to double-digit differences between Democrats and Republicans in each successive wave of our surveys. We also see that Independents are consistently more likely to agree that electronic voting systems increase the potential for fraud than are Republicans. Independents are also typically more concerned about such fraud than are Democrats. We also see that, since 2004, more Republican voters than Democrats in our samples agree that electronic voting systems are more accurate; fewer Democrats and Independents agree with that perception. In 2006, 59 percent of Republicans thought electronic

TABLE 7.3
Electronic Voting Evaluations by Voter Partisanship, %

	Republican	Democrat	Independent	Rep-Dem	Rep-Ind
Electronic voting systems increase the potential for fraud					
August 2004	31.4	34.4	42.7	−3.0	−11.3
March 2005	34.4	48.4	38.2	−14.0	−3.8
January 2006	25.0	40.1	46.3	−15.1	−21.3
October 2006	30.1	40.5	45.7	−10.4	−15.6
Electronic voting systems are more accurate					
August 2004	41.9	43.2	40.4	−1.3	1.5
March 2005	42.9	35.0	42.5	7.9	0.4
January 2006	49.6	44.4	43.3	5.2	6.3
October 2006	58.8	41.0	37.2	17.8	21.6
Electronic voting systems make voting easier for people with disabilities					
August 2004	63.9	67.2	53.9	−3.3	10.0
March 2005	62.7	57.4	62.3	5.3	0.4
January 2006	61.0	66.1	60.5	−5.1	0.5
October 2006	62.7	58.1	56.9	4.6	5.8
Electronic voting systems are prone to unintentional failures					
August 2004	35.7	50.5	42.2	−14.8	−6.5
March 2005	40.0	53.6	41.6	−13.6	−1.6
January 2006	33.2	46.0	49.9	−12.8	−16.7
October 2006	36.3	50.1	48.1	−13.8	−11.8

voting systems were more accurate, compared to 41 percent of Democrats. We also see increased numbers of Independent voters agree that electronic voting systems are prone to unintentional failures. In addition, a majority of Democrats across three of the four surveys agree that electronic voting systems are prone to unintentional failures.

In table 7.4, we repeat the previous analysis but this time examining differences between white and African American voters. The assessments of white voters regarding electronic voting systems are relatively stable when we look at perceptions regarding fraud, accessibility, and unintentional failures. The only change is that we see a growing trend among white voters in their agreement that electronic voting systems are more accurate. The views of African Americans on the same questions are much less stable and are generally more negative toward electronic voting. When we examine the question of accuracy, we generally find that, between 2004 and 2006, fewer African American voters agree that electronic

TABLE 7.4
Electronic Voting Evaluations by Race,%

	White	Black	White-Black
Electronic voting systems increase the potential for fraud			
August 2004	33.5	31.8	1.7
March 2005	37.0	52.5	−15.5
January 2006	36.2	48.0	−11.8
October 2006	34.8	37.5	−2.7
Electronic voting systems are more accurate			
August 2004	41.0	34.2	6.8
March 2005	40.8	33.4	7.4
January 2006	44.7	41.6	3.1
October 2006	48.4	28.8	19.6
Electronic voting systems make voting easier for people with disabilities			
August 2004	60.1	75.3	−15.2
March 2005	59.8	62.0	−2.2
January 2006	59.7	81.3	−21.6
October 2006	59.0	64.8	−5.8
Electronic voting systems are prone to unintentional failures			
August 2004	43.3	52.5	−9.2
March 2005	42.7	56.1	−13.4
January 2006	41.9	53.1	−11.2
October 2006	42.8	35.1	7.7

voting is more accurate. More African American voters have become concerned about the potential for fraud regarding electronic voting systems between 2004 and 2006, as well. One area where African Americans are more positive about electronic voting than are whites is on the issue of accessibility. African American voters strongly agree that electronic voting systems make voting easier for people with disabilities.

Our examination of the diversity of e-voting evaluations shows that there clearly is diversity of opinion in America about e-voting systems. Some of this diversity is understandable but some of it is also quite disturbing. As we see systematic differences of opinion by both partisanship and race, it is clear that the issue of electronic voting is a political issue. Despite the outcome of the 2006 election (both technologically and politically), it is unclear to us whether these racial and partisan differences will necessarily disappear in the immediate future. If differences based in party and race persist, or even deepen, they might lead to further controversy over electronic voting systems. These differences could ultimately lead to a situation where some groups of voters lack confidence in the utility of these particular voting systems. On the other hand, if election officials can successfully manage the continued implementation of electronic voting systems without further crises, if they can educate voters about their relative merits, *and* if voters come to believe that these new voting systems will record their vote accurately, then perhaps these differences will become less salient in the future. Today, it is clear that much work needs to be done to educate the public about electronic voting systems, and election officials and electronic voting machine vendors have much work to do to increase the confidence of some of their customers about the reliability of voting technologies.

DIVERSITY OF OPINIONS ABOUT THE PAPER TRAIL?

Now that we have examined public opinion toward e-voting generally, we can explore the same diversity of opinions about the various VVPAT alternatives we mentioned earlier in this chapter. We are again interested in exploring gaps in confidence and perceptions among specific segments of the voting population. In tables 7.5 and 7.6 we provide the breakdown of opinions about the VVPAT, based on the three questions in our second survey wave, again by voter partisanship and race.

We begin in table 7.5 with the opinions about the various VVPAT options in our survey data by voter partisanship. Here we find that, when asked about the choice between a paper or electronic ballot, there are interesting partisan differences: Republicans are somewhat more likely to desire an electronic ballot (nearly 53 percent stated that option), as are Independents (53 percent preferring an electronic ballot). But Democrats want paper ballots, with 64 percent of them preferring that option, and only 34 percent wanting an electronic ballot. Yet, when asked about whether they want a paper audit trail were they to vote electronically, we see strong support for the concept of a paper audit trail for an electronic ballot across all of the three different partisan groups. Finally, we see heterogeneity when voters were asked about whether they wanted to keep their paper receipt: 51 percent of Republican voters wanted to leave their paper receipt, but 46 percent of Democratic and 48 percent of Independent voters wanted to take their paper receipt with them when they went home.

Table 7.6 gives the same presentation of data, but for white and African American voters. Again, we see racial differences in assessments of the VVPAT. First, white voters are evenly divided about whether they prefer a paper or electronic ballot. By contrast, African American voters are much more inclined toward a paper ballot, with 56 percent stating a desire for a

TABLE 7.5
VVPAT Opinions by Partisanship, %

	Republicans	Democrats	Independents
If you had a choice, would you rather vote using a paper ballot or an electronic ballot?			
Paper ballot	45.7	64.1	43.4
Electronic ballot	52.8	34.0	53.0
Do not know	1.6	1.6	3.6
Refused	0.0	0.3	0.0
If you had a choice, would you rather vote using an electronic ballot that has a paper audit trail or an electronic ballot without a paper audit trail?			
Electronic ballot with paper audit trail	82.6	84.3	77.4
Electronic ballot without paper audit trail	10.3	10.0	14.1
Do not know	7.0	5.7	8.0
Refused	0.0	0.0	0.6
If you were to vote using an electronic ballot, which would you most prefer?			
No paper audit trail	16.0	10.5	12.9
A paper receipt that you can take home	28.7	46.0	48.4
A paper receipt you leave	51.1	40.1	36.1
Do not know	4.0	3.3	2.0
Refused	0.4	0.0	0.6

paper ballot. Then, when we examined whether voters wanted a VVPAT were they to vote electronically, we found again that there was strong support from both white and African American voters for an e-ballot with a VVPAT. A total of 79 percent of whites supported having a VVPAT and 84 percent of African Americans desired this option. Finally, we did find substantial differences between white and African American voters regarding what should happen with the VVPAT: a plurality of white voters wanted to leave the paper receipt in the polling place (43 percent), but a majority of African Americans wanted to take the paper receipt home with them.

VOTER COMFORT WITH VARIOUS VOTING TECHNOLOGIES

In our second and third survey waves, we also asked voters about their comfort with the various types of voting systems: "Regardless of whether

TABLE 7.6
VVPAT Evaluations by Race

	Whites	Blacks
If you had a choice, would you rather vote using a paper ballot or an electronic ballot?		
Paper ballot	49.6	56.1
Electronic ballot	48.0	41.5
Do not know	2.2	2.4
Refused	0.03	0.0
If you had a choice, would you rather vote using an electronic ballot that has a paper audit trail or an electronic ballot without a paper audit trail?		
Electronic ballot with paper audit trail	78.5	83.8
Electronic ballot without paper audit trail	12.5	11.0
Do not know	8.6	5.2
Refused	0.04	0.0
If you were to vote using an electronic ballot, which would you most prefer?		
No paper audit trail	13.8	11.3
A paper receipt that you can take home	37.6	54.4
A paper receipt you leave	43.3	34.4
Do not know	4.9	0.0
Refused	0.4	0.0

or not you have voted in the past, which of the following ways to cast your vote are you most comfortable with? Would you say [list of voting system options]?" We provide the responses to this question, from the second and third waves of our survey project, in table 7.7. We also give in the final column of table 7.7 data on the use of these different voting systems expressed as the estimated percentage of registered voters potentially using each voting technology in the 2004 presidential election, taken from a report issued by Election Data Services (EDS).[20]

We see in table 7.7 that voters, before the 2006 midterm elections, expressed comfort mainly in electronic voting systems (a plurality in each survey wave stated being confident with electronic voting systems), or an optical scan voting system (nearly 30 percent in each of the survey waves). Many fewer voters expressed comfort using the two voting systems that are rapidly being phased out during this same period, punch cards and lever machines. However, nearly 25 percent of voters in each of these two survey samples did say they were comfortable using either punch card or lever machines to cast their ballots.

TABLE 7.7
Voter Comfort with Voting Technologies, %

	Wave 2	Wave 3	November 2004 Use
Regardless of whether or not you have voted in the past, which of the following ways to cast your vote are you most comfortable with?			
Electronic or DRE	31.0	35.5	29.2
Paper ballot—fill circle	29.7	29.1	35.6
Punch card	19.7	16.9	12.4
Lever machine	14.1	11.1	13.2
Absentee ballot	0.5	1.2	—
Computer/Internet	0.3	—	—
Mail-in ballot	0.6	0.3	—
None/all/other	0.8	1.5	9.7
Do not know	3.2	4.3	—
Refused	0.4	0.2	—

When we compare the expressed comfort with actual voting system use in the 2004 presidential election, we find remarkably similar numbers. Although slightly more registered voters in the 2006 EDS report were estimated to potentially have used optical scan voting systems (36 percent in the EDS report) than electronic voting systems (29 percent in the EDS report), the correspondence between the voter comfort results and actual voting system usage in 2004 indicates to us that voters simply may be most comfortable with the voting systems they have used in the past, including the voting devices like punch cards that were heavily criticized following the 2000 presidential election and electronic voting devices that have become controversial more recently.[21]

Furthermore, although we do not present complete tables of additional results here (for brevity's sake), when we examine these same surveys' questions by important demographic attributes like partisanship, we see some evidence that voter comfort depends on their partisanship. In the third-wave survey responses for voters, for instance, we find that 39 percent of Republican voters express comfort with electronic voting devices relative to 35 percent of Democratic voters, and 33 percent of Independent voters. On the other hand, 26 percent of Republican voters expressed comfort with the use of optical scanning ballots, while 29 percent of Democratic voters and 31 percent of Independent voters said they would be comfortable using those types of ballots. Clearly, partisanship plays a role in determining what type of voting system voters feel comfortable with, in addition to the voting system they are experienced using.

VOTER EVALUATIONS OF VVPAT: A CASE STUDY

Before we conclude this chapter, we first note that sometimes our colleagues in election administration have the resources and inclinations to conduct small-scale pilot projects, to collect data and even to ask their customers (voters) about their evaluations of voting technologies. A case in point arose right as we were undertaking the final edits for the publication of this book, in the state of Georgia.

As readers have no doubt noticed, Georgia is a prominent example in our book, because of the aggressive and sweeping nature of its statewide implementation of electronic, touch screen voting immediately following the 2000 presidential election. Georgia continues to study how electronic voting works in the state, and in the 2006 general election, it implemented a unique pilot study: voters in three counties (Camden, Cobb, and Bibb) had the opportunity to cast ballots using a touch screen device equipped with a VVPAT. As part of the evaluation of the pilot project, the secretary of state's office worked with the University of Georgia's Survey Research Center to gather the opinions of voters, through an exit poll survey, about both the electronic voting machines and the VVPAT. Exit pollsters successfully interviewed 459 voters as they left polling places in these three counties on November 7, 2006, and their results provide an interesting case study.[22]

Generally speaking, it seemed that Georgia voters who used the VVPAT device in the 2006 general election seemed satisfied with it: 5.1 percent reported not being confident in the paper trail voting system, and more than 82 percent stated that Georgia should add the paper trail to all of its electronic voting devices. An amazing 97.9 percent of voters said that their overall voting experience using the VVPAT-equipped voting machines was good or fair, and almost 87 percent of voters said that they are very or somewhat confident in the accuracy and security of the touch screen voting machines used in Georgia. In comparing the security and accuracy of current DREs without paper trails to DREs with paper trails, the VVPAT added 2.5 percentage points to the confidence of voters.

Even though most voters provided overall positive evaluations of the VVPAT and the electronic voting systems, there were some issues that the exit poll survey revealed. Some voters did have trouble with the VVPAT: 3.5 percent reported some problem printing the paper trail (typically paper jams or the length of time it took for the paper trail to print), and 29 percent stated that it took longer to vote with the paper trail device. Furthermore, in a pattern much like that which we have discussed already in this chapter, nonwhite voters (56 percent of the exit poll sample was African American, 43 percent white, 0.7 percent Asian or Pacific Islander, and 0.2 percent Latino) were more likely to lack confidence in both the electronic voting system and the VVPAT. Nearly 6 percent of white voters said that they lacked confidence in the accuracy and security of the e-voting devices, relative to 12.9 percent of nonwhite voters. And when asked about the confidence of the paper trail voting system, 2.2 percent of whites said they were not confident, while almost 7 percent of nonwhites expressed a lack of confidence in the paper trail voting system.

No doubt, much more will be learned from the Georgia pilot projects as more information is made available to the public from the exit poll and from other data collected during the project.[23] But for our purposes here, this case study helps us make four points and confirms some of our results from our national surveys. First, when it comes to a real, hands-on experience with electronic voting technology, in a state which experienced significant problems in the 2000 presidential election, Georgia's voters seem generally satisfied with electronic voting's security and accuracy. Second, those who had the chance in this pilot project to cast a ballot using the VVPAT device also were generally positive in their reaction to the addition of the paper trail to the electronic voting system. But third, while the general evaluations of e-voting and the VVPAT in these three counties were positive, the exit poll data indicate that there were racial differences in confidence that need to be addressed (differences that parallel many found in our national surveys, presented earlier in this chapter and in our other research using our surveys on voter confidence and voting technologies).

Finally, our fourth point regarding the Georgia 2006 pilot project is that this is exactly the sort of work that election officials should conduct routinely. As we argued in our 2004 book on Internet voting, when election officials consider significant changes to their voting procedures or technologies, they should start small and pilot-test these changes first. Small, well-evaluated pilot projects can provide critical information for policy makers as they consider whether to scale up the new procedures or technologies to cover their entire jurisdiction. As part of these projects, furthermore, election officials need to do more to study customer satisfaction—they need to gather high-quality, detailed, and useful data on what their customers think about the services that are currently being provided, as well as what their customers think about the prospects of new services or technologies. Just like any major business, election officials need to think about the perceptions and opinions of their marketplace before making major changes in how they do business with their customers.

Summary

Recently, a study by the National Academies of Science (2005) on electronic voting (on which Alvarez was a member) placed strong emphasis on the importance of maintaining public confidence in the electoral process. This report noted that "Democracies derive their legitimacy from elections that the people collectively can trust. . . . Absent legitimacy, democratic government, which is derived from the will of the people, has no mandate to govern" (Celeste, Thornburgh, and Lin 2006, 2-1). Later, the same report also argues that:

> Election officials have been very concerned that various election problems in recent election years (most particularly in 2000, and in a lesser extent in 2002 and 2004) have shaken public confidence in elections, with the likely impact of depressing voter turnout in the short term and potentially undermining the legitimacy of government in the longer term. They have further believed that the controversy over electronic voting could have a negative effect in this regard in jurisdictions that use electronic voting, a point of particular significance when margins of electoral victory are very small. (2006, page 6-2).

The report then recommends three research items:

1. Identify the factors that influence public confidence in elections.
2. Study how confidence in and knowledge about elections and voting mechanisms varies across demographic groups.
3. Examine the impact on voter confidence of giving independent observers the ability to audit or scrutinize the conduct of elections.

In this chapter we have begun to provide some basic research into how Americans view electronic voting and the debate about the paper audit trail and have attempted to analyze how those opinions vary across demographic and political groups. We have identified a series of interesting results in our initial examination of the opinions of Americans and have undertaken what we see as one of the first major studies to attempt to answer the types of questions that need to be answered about the views of Americans about voting technologies. One clear result that we found early in this chapter is that Americans have a decidedly mixed view of electronic voting. On the four dimensions of electronic voting that we examined, only one—that e-voting systems are more accessible—received anything near majority support in our surveys. The other dimensions, both pro and con, were not ones that many Americans could clearly agree upon. In fact, we always found that sizable fractions of the population, usually around a third of our survey respondents, had no opinion at all about e-voting, indicating a relatively high degree of uncertainty, ambivalence, or lack of interest in these debates.

As to the basic question of the paper audit trail for electronic voting, we also found agreement in one place. If given a choice, our respondents would favor a paper audit trail for electronic voting machines over no paper audit trail at all. Where we found disagreement was over what to do with that paper audit trail. About 45 percent of our respondents favored a paper receipt they could take home with them but 37 percent favored leaving the paper receipt in the precinct. Opinion is decidedly in favor of a paper audit trail for electronic voting machines, but Americans are of two minds about how that paper audit trail should be used. Given the disagreement in the public on the form and use of the voter-verification systems, it is important to ensure that efforts to address the auditability problem do not lock election officials and vendors into one specific technology (the voter-verified paper audit trail as currently designed) and stop innovation into new and more interesting solutions to the auditability problem.

As to the demographic variation in these evaluations and opinions, we found two primary places where evaluations and opinions differ. One of these sources of variation arises from racial differences in the assessments of electronic voting and about the use of the VVPAT. We found generally that fewer African American voters agreed that electronic voting systems are more accurate and that larger percentages of African American voters agree that electronic voting technologies increase the potential for fraud. We generally found, though, that African American voters agree that electronic voting systems make voting easier for people with disabilities.

The dimension on which we constantly found divergent opinions, which we are also concerned about, is partisanship. We consistently found that Democrats are more skeptical of electronic voting technology, and Republicans are more favorable to electronic voting technology. Although we do not have sufficient data to test this hypothesis, we suspect that this partisan polarization on this issue arises from the partisan polarization of the debate over election integrity, with some Democrats using the issue of election integrity to mobilize supporters and keep the flames of the 2000 election debacle fueled.

We find this partisan gap disconcerting because partisan polarization about the process and procedures of elections may be very difficult to overcome, especially in the short term. Partisan debates over voting technology can be loud and rancorous, emotional, and devoid of facts and science. Passions can overcome rational debate, and in such an environment, it might be very difficult to persuade supporters of one camp or another that the process is producing election results that can be trusted—the very heart of the normative concern about the legitimacy of our democracy that serves as the foundation for this chapter.

Finally, we return to the example of Gilead, the pharmaceutical company we discussed at the beginning of the chapter, and to our case study of the recent Georgia pilot project on VVPAT. As a rule, it is safe to assume that major corporations invest in a great deal of resources on understanding their market, especially understanding the needs and opinions of their customers. Although we do not have systematic data on this point, we do feel safe in asserting that projects like the one Georgia recently undertook—especially projects that involve marketing surveys to assess the opinions of voters—are not the norm in the business of election administration. Obviously, when it comes to the related questions of electronic voting and the VVPAT, election officials and policy makers can learn a lot by simply asking what voters think of these technologies. They can learn what reforms might be favored by their customers, and how these reforms might best be marketed and sold to their customers. They can also learn more about which segments of their customer base may not be sold on the reforms, and use that knowledge to develop a strategy to better appeal to those customers in the future. As we discuss in the next chapter, such data are also critical for quality management in elections.

Chapter 8

A NEW PARADIGM FOR ASSESSING VOTING TECHNOLOGIES

The current debate about electronic voting and voting technology generally is hindered by the lack of a clear and coherent risk assessment model. Critics of electronic voting argue that direct recording equipment (DRE) is too risky to be used, but these critics also fail to provide a baseline level of risk from which to evaluate electronic voting. In short, they fail to address the question, Riskier than what? or to subject the baseline technology—paper-based voting—to the same scrutiny. Critics simply assert that there is a risk to voting on DREs but do not provide real assessment of the magnitude or likelihood of these risks occurring. We can imagine a simple comparison of system vulnerabilities between paper-based and electronic voting systems along a range of criteria. The 2000 election in Florida would be one example of the worst-case scenario illustrating the vulnerabilities of paper-based voting: ballots exist that cannot be counted because voter intent cannot be discerned, there are obvious issues with ballot design, and the result changes the outcome of the entire election. The 2006 election in the Thirteenth Congressional District in Florida would be a worse-case scenario for electronic voting, where ballot design seems to have led to a very high residual vote rate in an important race.

Critics of electronic voting also often fail to consider these risks within a real-world context and consider strategies that might mitigate or eliminate many of the risks they identify as potentially problematic. Even worse, many of the solutions that have been proposed to solve the problems associated with electronic voting have not been forced to undergo the same sort of rigorous threat-risk scrutiny to which the original DREs were subjected. For example, the threats of using VVPAT systems, which involve adding software and printers to DREs, add additional layers of complexity to electronic voting and introduce new threats to the use of these electronic voting systems, such as the opportunity for these systems to be sabotaged by intentionally jamming the printers. Other solutions, like parallel monitoring, which involves pulling equipment from precincts and testing it throughout the election to ensure that it is operating without malfunction, introduce mitigation without adding new threats.

Fortunately, risk assessment is something that both government and business have to address regularly, and there are models that can be employed to analyze the electronic voting case. In this chapter, we tackle how risk assessment models can be evaluated and how the current regulatory model for all voting technologies fails to provide effective oversight of voting systems. We rely on risk assessment models from the information technology sector and other sectors of the economy to appreciate how risk can be evaluated and mitigated. We also look to the General Accountability Office (GAO) for models of risk management, which considers how organizations should function in order to manage risks within its business operations. We then turn to the question of what a regulatory model should look like in the context of the increased use of information technology in elections. The current regulatory scheme for voting systems was designed almost two decades ago and has never been reengineered and reconceptualized to reflect the radically altered new world of election administration. Throughout this analysis, we attempt to suggest that all voting technologies should be analyzed equally and subject to equal scrutiny in the regulatory process. Given the problems that have occurred with all voting technologies—paper and electronic— one technology should not be given a free pass when the other is subject to a version of the precautionary principle.

ASSESSING RISKS

There are many risk assessment models, but they all tend to include similar components: developing the scope of the project in question, identification of potential risks and their likelihood, assessment of risks and develop- ment of priority risks, identification of mitigation strategies, collection of forensic data on implementation, and updating of threat-risk assessment on the basis of the forensics and changes in the implementation environ- ment.[1] The goal of any risk model is to determine what the risks are, how likely they are, and how either to stop the problem or minimize any poten- tial damage. It is helpful to consider each of these items to appreciate their importance in comprehensive risk assessment.

> *Project scope.* To engage in risk assessment, you have to understand the bounds of the project or activity you are evaluating. For elections, this re- quires carefully thinking through the complete set of activities that are a part of the voting process. For example, with electronic voting the scope might begin when the hardware and software are created and end with the system being stored after an election. For paper-based voting, it would ex- tend from ballot printing and the security of transmitting ballots from the printer to storing ballots after the election.

Identification of risks and likelihood. This is a two-stage process. First, every risk has to be identified. Second, the likelihood of the risk has to be quantified in terms of how much damage would occur if the risk was actualized and what the likelihood is that this actualization will occur. Combining damage and likelihood allows for each risk to be scored.

Assessment of risks and prioritization. By using these risk scores, each risk can be assessed and prioritized. This allows for each risk to be given appropriate consideration within the context of all known risks.

Mitigation strategies. For each risk, a mitigation strategy can be identified. Then, the costs of each mitigation strategy can be assessed against the risk score to determine the most cost-effective means of making the system secure.

Collection of forensic data. Forensic data are important for revising the risk assessment and mitigation strategies. For example, data regarding voting anomalies may result in the identification of new risks, but an analysis of polling place operations may suggest that other risks are overstated.

Updating model. The model is then updated on an ongoing basis in response to changes in the threat and risk environment.

Another model of risk assessment comes from the area of homeland security and terrorism prevention, being designed by the scientists at the Sandia National Laboratories. This model, the Vulnerability Assessment Model (VAM), "is a systematic, risk-based approach where risk is a function of the severity of consequences of an undesired event, the likelihood of adversary attack, and the likelihood of adversary success in causing the undesired event."[2] This methodology was designed for attacks on chemical facilities, and it is assumed that facilities cannot control their adversaries. Instead, the goal of the facility is to make itself a less attractive target for attack. This is accomplished by identifying the threats to the facility and creating a layered security system that can thwart the attack. One key to the VAM is the process of conducting an evaluation of system facilities and processes. This is done through creation of a process flow diagram and analysis of the conditions in which the operations occur and the policies and procedures that guide these activities. This evaluation allows the security threats and threat assessment to be mapped against actual practice at the facility.

Once risks are identified, they can be ranked by likelihood and severity and the highest-ranking items can be addressed first. The goal of the security mitigation is to allow for detection (alarms go off when intrusions occur), delay (it takes a long time to conduct an attack given the security), and response (attackers are intercepted at some point before the attack is complete). As the Sandia report emphasizes, "a well designed [physical protection system] will include . . . protection-in-depth, minimum consequence of component failure, and balanced protection."[3] In this system, protection-in-depth means that an adversary has to overcome multiple layers of security in order to cause problems with the system. The minimum consequence of component failure means that even if the system is penetrated at one point, the problem can be isolated so that it does not affect other parts of the system. The balanced protection concept means that, regardless of where the system is attacked, the level of protection-in-depth is the same. Once the system is in place, it can be tested in simulations to determine how well the security holds. Based on the results of these simulations, the security plan can be adapted to improve it.

The VAM methodology is intriguing as a model for elections for several reasons. First, it recognizes that attackers have choices and smart attackers will choose the weakest target to attack after engaging in some comparative evaluation of potential targets. Someone wanting to attack an election would likely do the same—find the weakest link in the weakest jurisdiction. Second, it recognizes that the characteristics of the facility where the chemical is being produced are critical determinants of system vulnerability. Likewise, studies of election security show that the security of the system deployment—the security of the election management software, the way in which voting machines are deployed in the precinct, and the way in which absentee ballots are handled and processed—all affect the security of the voting system. Third, VAM recognizes that there is no "silver bullet" to security but instead security must be layered and balanced. Finally, VAM is predicated on testing and revising the security of a system on an ongoing basis; this is not a one-shot evaluation process.

One of the problems with the current debate about electronic voting is that there has been little effort to engage in full and complete risk assessment by many participants. This has been especially true in the area of electronic voting, where academic researchers have often failed to compare electronic systems with a benchmark, which would allow for the threat of moving to an electronic system to be compared to the risks associated with staying with the status quo (typically paper-based and in some cases by-mail voting systems). In most cases, the paper system is considered by critics of electronic voting to be a "gold standard" for security, even though America has more than 200 years of experience in vote fraud committed with paper ballots. Rarely is the security of paper-based systems evaluated with anywhere near the same scrutiny of electronic systems (cf., Shamos 2004).

Two interesting examples of the weakness of the assessments made of electronic voting are the analyses that were done in 2003 and 2004 of the SERVE Internet voting system and the Diebold AccuVote-TS. What is interesting about both of these is that the analysis conducted addressed only the identification and, in one case, the attempted ranking of risk. What did not happen was an effort to address the first step—defining the project scope—or the later step of examining the threats and risks once a risk mitigation strategy was put into place.

In the case of SERVE, the critique of the project failed to consider that it was a pilot test—not a national implementation of Internet voting. It also failed to consider the comparison to the paper-based UOCAVA voting process that currently existed. Finally, it failed to take into account that, even though the project would have taken place in only a few handfuls of jurisdictions, even a small number of jurisdictions introduced a variety of layers that would have made many of the attacks difficult (for example, each jurisdiction would have had different ballots and ballot styles, introducing ballot complexity) and would have introduced many additional observers in the process who would have been in a good position to detect and identify attempts to infiltrate or attack the system.

As to the evaluation of the Diebold AccuVote-TS system, its weaknesses are illustrated in the studies that were conducted after the initial Johns Hopkins/Rice study was released. In both cases, the later evaluators noted that the Diebold system did have weaknesses, but the academic analysis had failed to consider how these systems were implemented in polling places and how basic mitigation strategies could address the problems identified in the initial report. In short, the initial study was not a threat-risk assessment but simply the identification of risks without completing the process. For example, the trusted-agent report prepared by RABA Technologies LLC states in its preliminary findings that "even a cursory examination of [the Johns Hopkins/Rice and SAIC] documents (and most of the subsequent public debate) indicated that the analyses were undertaken with less than full knowledge of the technical, operational, and procedural components that must be considered together in assessing any voting system". The report notes that the Johns Hopkins/Rice report is a "thorough, independent review of the Accu-Vote-TS source code" but goes on to state that, "just as the authors correctly point out that there are standards by which the development of a secure system may benefit, they chose to disregard the similar standards by which systems must be evaluated."[4]

In short, the Hopkins/Rice evaluation did not consider the voting system as a system or as a part of an overall election administration process and did not consider the security of the system in total but rather in parts. To consider an engineering metaphor, the appropriate way of evaluating a structure is not to note that a bolt might rust and break but is to consider how the bridge reacts and isolates such an event. Likewise, the security of a voting machine needs to be considered in regards to how it isolates and reacts to failures.

But in all fairness, even though these evaluations may not have been as comprehensive as we argue they must be in the future, the fact remains that all three analyses of the Diebold AccuVote-TS found security shortfalls in the system. These problems were not discovered in the voting system certification process because the Independent Testing Authorities are charged solely with evaluating the functionality and construction of a voting system; evaluation of system security or threats and risks to the system are not considered by the authorities because it is outside the scope of their charge. This failure in the process is something we take up later in the chapter.

A second problem with the current debate over the security of voting is that a "magic bullet," the VVPAT, has been introduced into the debate as a means of ending the discussion over security. Although some critics would like to see electronic voting terminated entirely and all voting done on paper ballots that are counted using precinct-based optical scan systems, critics of electronic voting argue that VVPAT will make electronic voting secure because it produces a paper ballot that can be audited.[5] Like paper systems, are however, the VVPAT is also never forced to undergo a threat-risk assessment to determine what, if any, new threats are introduced to the voting process when a paper printer is introduced to the voting process.

MANAGING RISK

Once risks are identified and a threat-risk assessment has been conducted, the findings of the process have to be implemented. Organizations do not implement risk assessment models in a vacuum. But typically implement their risk management practices within a specific business management and operational context. The GAO (1998) has found that, in the area of information security, leading organizations are guided by five principles:

1. Assess risk and determining needs.
2. Establish a central management focal point for information security.
3. Implement appropriate policies and related controls.
4. Promote awareness to educate users.
5. Monitor and evaluate policy and control effectiveness.

These five items are not static but are part of a risk management cycle. As Figure 8.1 shows, the central management focal point is the hub around which the other four items rotate. This hub consists of professionals who carry out the key information security activities. The GAO stresses that these individuals should be permanent staff who have direct access to the organization's senior executives. These are the people who have the job of educating staff, creating a culture that respects and values high security, and keeping senior executives aware of changes in the threat-risk

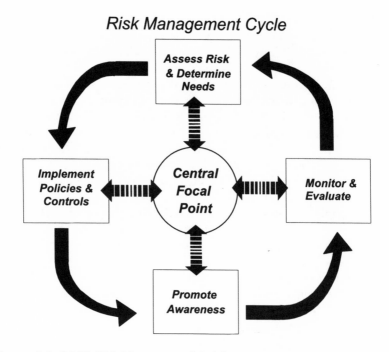

FIGURE 8.1 GAO's Risk Management Model
Source: GAO. 1998. Executive Guide: Information Security Management. May, GAO/AIMD-98-68

environment. Creating this type of information security requires resources and commitment by state and local election officials and rethinking the way in which elections and election operations are conceptualized. It also requires holding managers accountable for failures for managing risks and implementing mitigation strategies against known risks.

Each of the nodes in the risk management cycle generally correlates with the items we identified at the outset as being critical to conducting a threat-risk assessment. Risks are assessed and ranked, mitigation strategies are implemented and controls put into place, and then these risks are monitored. Election officials need to monitor the implementation of elections in the polling places and in their offices areas where data for risk mitigation are not commonly evaluated.

Election officials are often too busy managing their elections and are too understaffed to send out workers solely for the purpose of evaluating election operations at the polls. One example of a jurisdiction that does engage in this type of activity is Chatham County, Georgia (Savannah). Kathy Rogers, then the supervisor of elections in that county, implemented a system where experienced election workers who had been given extensive training were made roving election observers. These individuals would visit up to ten precincts at least three times during the day—in the morning, afternoon, and at closing—to determine how elections were being implemented in each place. These individuals not only evaluated problems, using an evaluation tool developed by the county, but were empowered to make changes at polling places to mitigate against problems that were observed. Once the system was in place, these county election observers were able know the strengths and weaknesses of each polling place they were visiting based on previous experience. The survey examined key issues in election security, although they were not conceptualized as such at the time. These issues included making sure that there was a clear flow of traffic in the polling place, that election workers were manning the ballot box (in this case, a precinct optical scan system), that the ballot box was located close to an electrical outlet (optical scan systems require power to scan ballots), and that the blank ballots were secure.[6] A system like the one used in Chatham County could be expanded to be a more security-oriented evaluation. This would ensure that the polling places were fully implementing the security strategies that were developed by either the state or local election jurisdiction and were doing so throughout the day.

Not only are evaluation and monitoring important, but so is educating people about security risks. In fact, the only difference is that the risk management cycle inserts promoting awareness between mitigation implementation and risk monitoring. In elections, promoting awareness is critical because security has not traditionally been the frame through which elections were viewed and implemented. In the poll worker training we have observed, security is one issue, but the primary issue is getting voters efficiently through the voting process without making errors and disenfranchising voters. With security becoming a larger issue—and elections generally becoming more highly evaluated and contested—promoting awareness will become even more important.

IMPLEMENTING A RISK MANAGEMENT MODEL

Is this type of risk management model for elections outside the realm of possibility? And what would an effective risk management system look like? One model for developing and implementing a risk management model can be found in Travis County, Texas. The Travis County model is predicated on the following idea: what would an election official have to do to defend his election in court under the federal rules of evidence? Under rules of evidence, you need to have three things: something physical, like a report or an audit log; details on the chain of custody of the document (who created it, when, and signatures acknowledging this), and secure storage of the physical item along with the chain of custody documentation. At a crime scene, investigators want to find the murder weapon, seal it in a bag and sign over the tape that seals the bag, and store it in the evidence safe. For election officials, evidence might include whether they handle absentee ballots, show a constant chain of custody, and store them securely.

Travis County has engaged in a risk assessment evaluation and then developed a set of security practices to address those risks.[7] This involved a three-stage process. First, the county identified all of the steps in an election. As Figure 8.2 shows, this starts with the acceptance testing of voting equipment and its dormant storage in a warehouse; moves through early voting, absentee voting, and election day voting operations; and ends with tabulation, postelection auditing, and then return to dormant storage. Second, after each operational step was studied, the county identified the period at which the system was at its highest level of vulnerability. This period ran from the deployment of equipment in early voting through the receipt of ballots and close of voting on election day. Third, the county developed mitigation strategies for each step. Examples of these strategies (shown in Figure 8.2) can be simple (when equipment is in storage, cover the equipment with plastic covers in case the sprinkler system is activated) or more complex (ensure the loaded software is the same as the certified software, or conduct a hash code test against the file that was sent by the independent testing authority to the National Institute of Standards and Technology National Software Reference Library). Still other strategies require quite a bit of planning and resources (on election day, randomly pull one precinct's voting equipment, send emergency equipment to that precinct, then conduct a parallel monitoring effort in a secure area with multiple cameras observing the process).

Officials in Travis County then applied the three-part evidence standard to their operations. For example, during early voting, law enforcement officials transport the electronic ballot boxes to a location where they are stored in a secure room with a surveillance camera and then guarded by law enforcement. Each day, law enforcement transports the electronic ballot boxes back to the early voting sites. At each step, individuals are signing an audit log and much of the time are under electronic surveillance—showing the maintenance of the chain of custody—and the electronic ballot boxes are being stored in a secure location (analogous to an evidence locker).

Much of the security also comes from a combination of physical security and multiple signoff requirements. For example, ballot software is protected by physical security because only five employees know the pass code to the room where the ballot programming and tabulation

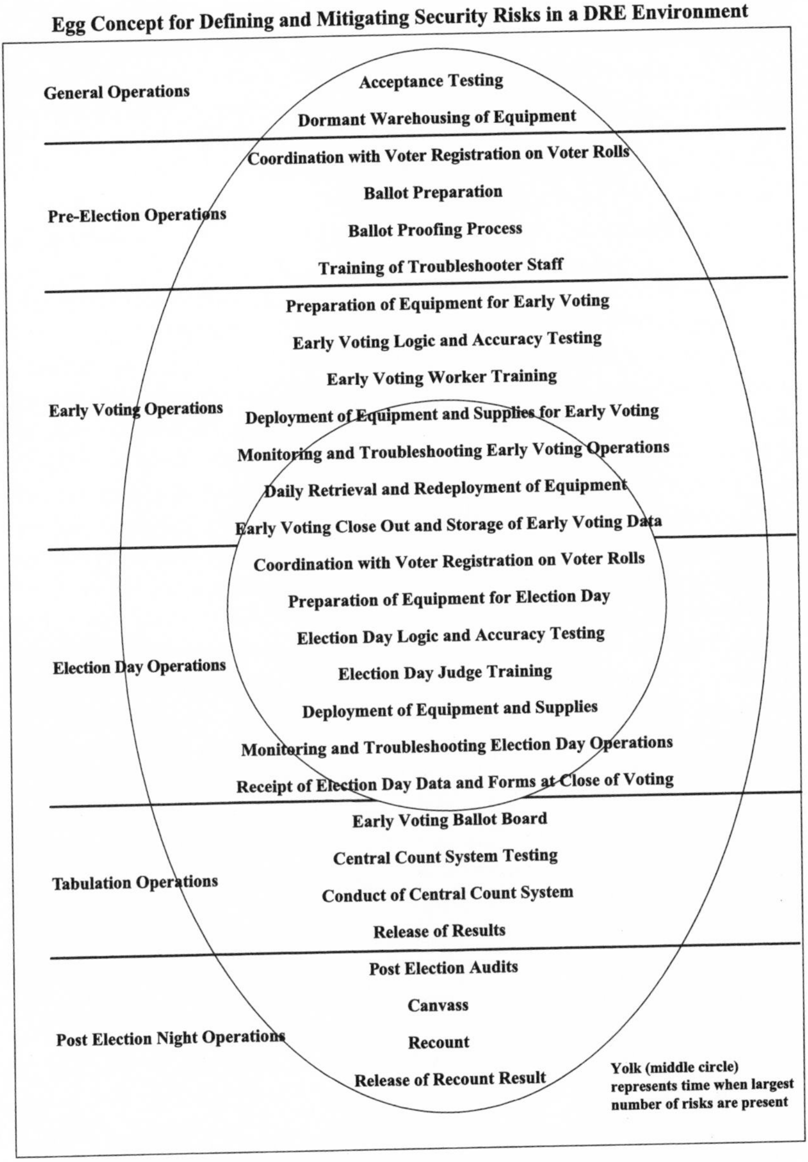

Egg Concept for Defining and Mitigating Security Risks in a DRE Environment

General Operations

Acceptance Testing

Dormant Warehousing of Equipment

Pre-Election Operations

Coordination with Voter Registration on Voter Rolls

Ballot Preparation

Ballot Proofing Process

Training of Troubleshooter Staff

Early Voting Operations

Preparation of Equipment for Early Voting

Early Voting Logic and Accuracy Testing

Early Voting Worker Training

Deployment of Equipment and Supplies for Early Voting

Monitoring and Troubleshooting Early Voting Operations

Daily Retrieval and Redeployment of Equipment

Early Voting Close Out and Storage of Early Voting Data

Election Day Operations

Coordination with Voter Registration on Voter Rolls

Preparation of Equipment for Election Day

Election Day Logic and Accuracy Testing

Election Day Judge Training

Deployment of Equipment and Supplies

Monitoring and Troubleshooting Election Day Operations

Receipt of Election Day Data and Forms at Close of Voting

Tabulation Operations

Early Voting Ballot Board

Central Count System Testing

Conduct of Central Count System

Release of Results

Post Election Night Operations

Post Election Audits

Canvass

Recount

Release of Recount Result

Yolk (middle circle) represents time when largest number of risks are present

FIGURE 8.2 Travis County Election Process and Risk Assessment
Source: See Dana DeBeauvoir, "Method for Developing Security Procedures in a DRE Environment"

system is housed, the computer produces a contemporaneous audit log, and multiple signoffs are needed for a ballot to be implemented in an election. In other cases, the county cannot even access components of the election without the sheriff's department being present because the sheriff has the final key needed to open the innermost secure area.

Similar security measures are taken to safeguard paper absentee ballots in Travis County. For example, the outer envelope for returned absentee ballots is removed in a separate building in case someone attempts to disrupt the election by placing a hazardous substance in a returned envelope; imagine how a single anthrax letter could completely contaminate a local election office. The inner envelope containing the absentee ballot is taken to a separate building where determination is made whether it should be included in the final canvass. The work of evaluating the envelopes, separating the outside envelope from the privacy envelope, and then separating the ballot from the privacy envelope, is done by different teams at different tables, but all under the watchful eye of several surveillance cameras. Again, these paper ballots are placed under unique separation security—the sheriff secures the ballots in a room for which he has the only key within a secure section of the local election office that only the election staff can access via coded keypad.

One critical facet of Travis County's risk model is that officials take ballot security seriously across all voting platforms; they treat the threat that paper ballots will be tampered with or stolen just as seriously as they do the possibility that electronic ballots will be tampered with and stolen. It is no surprise that the Travis County threat assessment process has been publicized as a model for other election officials to study and follow.[8] The risk model in Travis County also raises a number of fundamental questions. First, is the security assessment is sufficient? Second, how can other election jurisdictions develop, refine, and implement their own threat assessment plans, especially given their limited resources? Third, how do we regulate the security of the ballot so that all voters can be confident that their ballot is counted without problem? Only by developing a stronger state and national regulatory model for all voting systems can we be in a position to answer these questions.

A BETTER REGULATORY MODEL FOR VOTING SYSTEMS

When examining the current system of regulating voting technology, we can imagine a range of options. On one end of the scale there would be no standards process; however, a close second is a completely voluntary standards process. Standards in this setting serve as a marker, a guide that can be used if any entity decides to require voting systems to meet some specific standard. Once we agree on the need for standards, the key questions are: should the standards be mandatory, who should enforce the standards, who should be the target of the standards, and are the standards complete for the purpose and goal of the project?

Given that we already have standards for certain aspects of elections, the key question in the current election environment becomes: what does a regulatory model for voting systems look like? Few people would argue with the premise that the current regulatory model for voting systems has several flaws:

- The current Voluntary Voting System Guidelines are called voluntary and guidelines for a reason; there is no federal requirement that the voting systems that are used in local jurisdictions comply with the existing standards. Some states do, however, require voting systems used in the state to comply with the guidelines. Although Section 301 of Help America Vote Act (HAVA) does establish some baseline standards for voting systems purchased with HAVA funds, there is no federal requirement that these systems meet all of the Voluntary Voting System Guidelines.
- The system is generally static; once certified, a system is forever certified, although recent decisions in California by two successive secretaries of state have started to change this dynamic. However, even in California there is no systematic process for reviewing voting systems in light of new data, new technologies, or new circumstances to renew a certification.
- Even if a voting system goes through the certification process, the process is limited in scope and focuses primarily on system functionalities, although the draft 2005 Voluntary Voting System Guidelines begin to delve into the areas of accessibility and security.
- The process by which the Independent Testing Authorities evaluate voting systems is opaque. The public and election officials alike are limited in their ability to get good information about the issues that were discovered through the testing process because the process is an arrangement between the vendor and the authorities.
- Voting systems are currently evaluated completely outside the environment in which they will be implemented. There is no requirement that the election jurisdiction implementing the system show any capacity to be able to handle the implementation and ongoing use of the new voting technology.

There are several ways of addressing the problems with the current regulatory environment. The first is to try to layer an external oversight process onto the existing system. This is in essence what is being done today with both VVPAT and parallel monitoring. One problem with the current system is that it is generally applied differentially to one class of voting equipment, DREs, when the threat that is being addressed by both processes are threats to all vote tabulation systems. Therefore, these techniques could be applied to *all* vote tabulation systems, not just DREs. Both of these processes are predicated on the idea that no amount of regulation can determine if an electronic tabulation system is working properly. Therefore, it is necessary to provide real-time auditing for the system. VVPAT allows voters to review a paper version of their ballot to see if it matches what is on the DRE screen and then cast their vote if the two match.

Parallel monitoring is almost real-time monitoring of elections. It involves randomly selecting a set of voting machines that were to have been deployed on election day—standard replacement equipment is sent to the precinct instead—and that equipment is then secured in a room and placed under video monitoring and the machine is voted all day, as it would be under normal election conditions. The machine totals are then compared to the ballots that were voted. The rationale for this process is that, were anyone to tamper with the voting system software so that it created anomalous results—for example, turned every tenth vote for a Republican for president into a vote for a Democrat for president—this anomaly would be noted in the results from the parallel monitoring. The random selection of voting equipment and testing of the system under real voting conditions is designed to ensure that the voting equipment cannot "know," through some logic functionality, that it is being tested and not being used in a real election.

Parallel monitoring has been used extensively in California since 2004, and reports on these efforts are available from the California secretary of state. We used some of the data from the California parallel monitoring effort in Chapter 6, on the performance of electronic voting systems in the recent election cycle. To explain how this process works, as well as to note how the errors that have occurred with DREs in the parallel testing process have all been human-factors-related (data entry errors) and to emphasize that there was no evidence of any problems with the DREs themselves, we quote from the 2004 parallel monitoring report:

> Test scripts served as the tool to achieve the main goal of validating the accuracy of the DRE equipment. Test scripts were designed to mimic the actual voter experience. Each script represented the attributes of a voter (party affiliation, language choice) and specified a candidate for which the tester should vote in a specific contest. The test script form was laid out to record requisite details of the voting process for a "test voter" and served as a means to tally test votes and assist in verifying if all votes were properly recorded, summarized, and reported by the DRE unit. For each county, 101 test scripts were developed. All contests, contest participants, voter demographics, script layouts and contents, and monitoring results were entered into a MS

Access(tm) database. The database was a tool to manage 242 contests, over 1,000 contest participants and approximately 52,000 test voter selections from over 1,000 test scripts. The database also served as a tool to verify the accuracy and completeness of the test scripts. . . .

Test teams were scheduled to arrive at their assigned county at varied times on the morning of November 2, 2004, to meet with county representatives, retrieve the voting equipment from storage, and be escorted to the testing room. Test teams followed a specific test schedule that identified set times for executing the 101 test scripts on each DRE unit. The schedule provided for 9.25 hours of testing over a 13-hour period. All testing activity was video recorded. . . .

At the completion of the testing, teams produced the closing tally report for their assigned DRE unit. The test teams did not reconcile the tally tapes in the field and had no knowledge of the expected outcomes. The analysis of the data and the reconciliation of actual to expected results began on November 3, 2004. The analysis included a review of the discrepancy reports for all counties and the videotapes, as necessary, to determine the source of all discrepancies. Results of the reconciliation analysis indicate that the DRE equipment tested on November 2, 2004, recorded the votes as cast with 100% accuracy.[9]

Other models provide similar controls over the quality of election outcomes. For example, there are ways to allow voters to ensure that their vote was included in the final election results and tabulated accurately. Several corporations, including VoteHere and Scytl, have produced systems that provide for electronic cryptographic voter verification that voters can take home with them. Some of these systems also provide for encrypted and redundant storage of the electronic votes. The process these systems use is simple: ballot options the voter has selected are encrypted, similar to the way your password is encrypted when you log on to a secure Web site. The encryption process produces a code, for example, the alpha-numeric combination R1M2A3, and this code is shown on the screen and printed out for the voter. The ballot is then cast. The key to the process is this: if the ballot is tampered with by the voting machine when it stores the vote choices, a different alpha-numeric combination will be produced for that ballot. So if the ballot produces R1M2A3 when it is decrypted for tabulation, the ballot has not been tampered with. If it produces T9E8H7, then we know that there is a problem with the ballot. Some electronic verification systems would then allow you to retrieve the encrypted ballots from their independent storage unit and count the ballots as cast by the voter. These processes again are currently thought of as adding layers to the existing regulatory process without considering how the process itself works as a whole.

A COMPREHENSIVE MODEL OF VOTING SYSTEM REGULATION

Several comprehensive regulatory models can provide guidance for how to improve the regulation of voting technology. Perhaps the most comprehensive is the regulatory model used in the pharmaceutical industry. The Food and Drug Administration (FDA) model—outlined in the report *Pharmaceutical [Current Good Manufacturing Practices] for the 21st Century, A Risk-Based Approach*—not only determines whether a drug gets on the market, but also includes regulation of the manufacture of drugs, the performance of a drug over its lifecycle, ongoing implementation of scientific risk-based approaches, and the implementation of a quality systems framework for FDA activities in this area. We first consider the strengths and weaknesses of the FDA model, and then consider how such a model could be applied to elections.

Developing Internal Capacities

The first component to the FDA regulatory model is that the agency has a model for regulating their own management practices. Quality management is fundamental because just having a strong risk assessment process, or mitigation strategies that are not well managed, can minimize the effectiveness of these processes. At the FDA, the Quality Systems Framework allows the agency to improve product and service quality by having a process for capturing key information and then using that information to manage its resources and the life cycle of the drugs it regulates.[10] The Quality System Framework (QSF) begins with developing the system's infrastructure.[11] This involves defining and documenting the scope of the organization's internal work processes and the work product and services that will be managed under the QSF rubric. It also involves documenting the external or outsourced work processes, products, and services that will affect product quality.

With this documented process developed, the next step is to develop the strategic management—the leadership, policies, objectives, plans, and organization structures—that are needed to implement the QSF. The QSF has to be integrated into the overall organizational strategic plans and mission and be given priority by the organization's top leaders. Moreover, the entire organization needs to be trained why the QSF is the guiding management infrastructure. Once this is done, the organization can establish best-practice benchmarks, a monitoring process for evaluation, and place this system within the structure of the organization so they align.

Third, the organization has to develop a method for evaluating its internal lifecycle management process. This involves carefully planning its work, designing new services and processes, and evaluating that work to ensure that it is free of problems. This continuous improvement process requires multiple reviews of work that is done and continual evaluation of the products that are being produced. Finally, the entire system is subject to a quality system evaluation process that involves continually monitoring and measuring key system metrics. Then, these metrics are analyzed for problems and trends, which are addressed through an open process that involves both internal and external stakeholders. These findings feed into the management of the entire process; as problems and issues are identified, the QSF has to change to address them.

The Election Assistance Commission (EAC), the states, and local election officials could all benefit from having an internal process that documents their organization's work and that requires continual oversight and evaluation of its effectiveness. Such a system would serve to catch many problems that have occurred in elections, like the notorious butterfly ballot, the apparent ballot design failure in the 2006 congressional election in Sarasota, Florida, or the failures to send absentee ballots to voters in King County in 2002, before these events occurred. What such a system would require of election officials is the type of continuous monitoring and measurement of key election metrics that many are sorely lacking today. Implementing such a quality system for elections would have several requirements. First, state and local election officials would have to have a common definition of what constitutes a quality election and common definitions for various aspects of the election process. Second, they would have to collect data in every election for an array of variables that affect the quality of the election process, from training to security to voter satisfaction. Third, states and localities would have to have a plan for how to implement changes to the election process to address the problems that are identified in this quality management process.

To understand the issues associated with the data definition and data collection component of the process, consider the EAC *Election Day Survey 2004*, the federal government for the first time ever attempted to collect election administration data from all states and localities. The survey was difficult to implement because states have differing definitions of the same concept, such as what constitutes an absentee or early voting ballot. States and localities also do not collect the same data or collect data in uniform ways. Lack of data collection generally and lack of uniformity combine to make it difficult for either national or state election officials to conduct evaluations of quality or of problems in the election process. But the fact that at this basic level it is difficult to get election officials to retain such data, in a common format, and to provide it in a timely manner to the federal government points out how difficult it will be to get the thousands of local election jurisdictions throughout the nation to develop and use common quality management procedures.

We can also look to a simple question that is only now being examined by election officials and scholars: who are poll workers? There is very little basic research that has been done on poll workers, such as their average age, education level, comfort with technology, views of their training, and similar factors. If poll workers play the critical role in the implementation of election systems that anecdotal evidence suggests and the role in public confidence in the electoral process that some preliminary studies suggest,[12] then the lack of data on poll workers is a severe gap in our ability to understand election quality and to implement a system of metrics for knowing when elections are or are not being implemented at a high level of quality.

Developing a Rigorous Oversight Process

Although people generally focus on the FDA's work in determining if a drug can come onto the market or can stay on the market once approved, the other key regulatory component of this work is to evaluate the production of drugs as well. The manufacture of drugs is regulated in law and regulation, which defines Current Good Manufacturing Practices (CGMP), a term of art in the law. Unlike the Voluntary Voting System Guidelines, the CGMP are neither voluntary nor guidelines; instead, they are binding regulations that govern how drugs are to be manufactured. The FDA not only issues the CGMP, and guidance on how to implement it, but it also conducts inspections to ensure that CGMP standards are being followed. If they are not, the FDA can take punitive action against the violating corporation. The regulations and guidance are guided by the best science possible and effective risk-based processes that mitigate risk and still encourage innovation and the improvement of products and processes for producing these products. The goal is to have a flexible innovative industry that remains well regulated.

The manufacturing oversight process is but one component of drug oversight; the FDA is obviously very interested in also evaluating the human-drug interaction process. This process begins during the drug approval process, which involves extensive testing of the effectiveness of a drug and the externalities that are produced from the drug. (For example, a drug might be highly effective at treating cancer, for instance, but at the cost of causing paralysis.) These data are collected in clinical trials—which are conducted by the manufacturers using scientific standards and practices established by the FDA—and then evaluated by the FDA. In the case of elections, an analogous requirement might be that voting machine vendors test their equipment on specific voting populations under controlled settings so that system usability can be tested. This process could also require voting systems to undergo tests against security threats to the system before the system is certified. Of course, all voting systems would have to undergo these tests, not just specific classes of voting systems.

The other aspect of evaluation of drugs is to determine how well drugs work on an ongoing basis. This evaluation requires the collection of both positive and negative outcome data related to the drug. Adverse reactions are reported and evaluated by the FDA, and drugs that have a high rate of adverse reactions in real-world use may be pulled off of the market. The Vioxx debate is one interesting example of this. Vioxx was initially approved to be used at a certain dose over a specified time frame. However, the drug was used in practice for longer time frames and at different dosages. The same can be true for voting technologies. A given specific voting system brand may work well in theory, but in practice problems may arise in implementation that make it a candidate for being removed from the market. To think back to the drug manufacturing side of the equation, drugs and medical devices also often come with specified life-spans. Drugs expire and medical devices may be known to require maintenance or replacement after a specific date. Again, a specific class of voting technology or brand of voting technology might also have a similar requirement for maintenance or replacement after a specific date.

The medical analogy allows us to consider what a similar regulatory scheme for elections would look like. Elections have two "manufacturing" components. First, there is the manufacturing of voting systems—from ballot design software to voting machines to vote tabulators and tabulation software. These systems have an obvious analogy to the FDA's CGMP system; voting machine manufacturers, voting equipment software companies, and ballot printers alike would be subject to oversight and regulation that would ensure that such systems are appropriately certified, manufactured properly, and meet standards for security and usability. Moreover, data on system problems would be collected and fed into a centralized system that would allow election officials and policy makers to know about issues associated with voting and voting systems—the equivalent of adverse reactions with drugs—and address them if appropriate.

The second half of the "manufacturing" process of elections is the actual implementation of the election in a state or locality. In many recent elections, the anomalies in the election have not been caused by voting technologies but instead are caused by issues of election administration. For example, questions regarding the allocation of voting machines and poll workers across jurisdictions or the design of ballots have been the cause of many questions and problems in recent elections. If states could collect this type of "manufacturing" data from localities, they would be in a position to understand where problems exist and to put debates over elections into the realm of fact, instead of mere speculation. States and localities would also be able potentially to "certify" that an election had been conducted in accordance with best practices.

Rigor and Uniformity in Regulatory Standards

One of the reasons why the public trusts the drug regulation process, even though there are occasional system failures, is that it is a scientific and rigorous process that is equitably applied. Quality data are collected and evaluated on an ongoing basis, and regulatory action is taken in response to those data and evaluation processes. There is also uniformity to the process; all drugs and devices are treated equally. This uniformity principle is one of the key problems with the current argument over the need for voter verification for DREs. The arguments for having an audit trail have existed for some time; this argument is made explicitly, for example, in the 2001 report of the Caltech/MIT Voting Technology Project. However, in the current debate, the argument is made only for a specific class of voting technologies: DREs. The rationale is that paper ballots can be recounted. However, a recent report by electionline.org examining state recount laws shows that this is in fact not the case; there are very limited situations in most states for holding a recount, so the ballots cannot just be recounted. Moreover, with a paper system, voters have the same problem of knowing if their vote was counted as cast, because an optical scanner does not inform voters how the ballot was tabulated. Therefore, there must be uniform requirements for auditability of tabulation systems that mirror the VVPAT requirements that have been articulated for electronic voting technologies. Similarly, precinct tabulators could be tested in a parallel monitoring process to ensure that there are not errors in the tabulation of ballots just like those that can theoretically occur with DREs.

The key question in regulating elections is in determining what, if any, components of the election process should be regulated at either the state or federal level. Currently, there is a limited federal regulatory framework for elections. The U.S. Department of Justice is rigorously enforcing various aspects of the Help America Vote Act and has historically enforced the language minority provisions of the Voting Rights Act. However, the voting process itself—the ballot design, voting systems and voting machines, usability of ballots or machines, election audits, what constitutes a vote on different voting systems, and the like—are not regulated at the federal level. The benefit of federal regulation is that it can create common understandings of the voting system and process. For example, there is no reason for voting equipment to meet different standards for usability or security across states or localities. There is also a strong argument that could be made for having specific requirements for the design of ballots, given the problems that have arisen related to ballot design in recent elections. However, other regulations can serve to make the voting system very brittle. The decentralization of election regulation does serve to isolate problems and allow states and localities to experiment. For example, the "vote center" innovation that has been used very successfully in Larimer County, Colorado, which has boosted turnout among low-propensity voters, would be disallowed if there were rigid federal rules for the size of precincts or allocations of voting equipment. In addition, given the variations in voting mode—among early, absentee, and precinct voters—across states and other related issues, state laws will continue to govern much election activity. The federal role may be to strongly encourage states to develop highly robust regulatory regimes to address elections within their own state legal framework, with fiscal support for this regulatory effort coming from federal coffers.

UNDERSTANDING THE CUSTOMER

As we argued in chapter 7, election officials can do much more to gather information on what their primary customers want from the election process, to try to implement procedures consistent to those expectations, and then to work continually to measure how well administrative implementation meets customer expectations. Readers familiar with quality management will see that these themes are similar to those in the widely used "Six Sigma" approach to quality management, where customer specifications define product defects, and managers attempt to impose quality control so as to minimize product defects (commonly to a threshold below 3.4 defects per million opportunities).[13]

Critical to Six Sigma—or any other type of quality-based management approach to election administration—is the imperative to understand the expectations of the product customer. No matter what management approach election administrators take, as they seek to implement new voting technologies (be they electronic or paper, or something in between) and new election procedures, they should follow a careful, research-based process. They need to get information, early in the decision process, from all key stakeholders in their jurisdiction, especially the voters, about what expectations exist for voting system performance. All the key stakeholders, again including voters, should be involved in all stages of the acquisition and implementation process. And once new voting systems are in place, the "voter-as-customer" needs to be queried continually about the voting system, to ensure that election jurisdiction continues to meet (or exceed) expectations.

Although much has been debated about whether new technologies are empowering citizens vis-à-vis their government, in our opinion the debates over voting technologies and election administration are not going to end soon. There are now many Web sites and blog sites devoted to discussion and debate about voting technology, there are many in the research community now actively involved in studying voting technology and election administration, and there are thousands of activists throughout the nation involved in this issue. Managing the expectations of the public will be a critical task for election administrators, so it is imperative that they begin to understand those expectations by talking with their primary customers, the voters. Gathering systematic data about voter expectations and evaluations is a critical component of any attempt by election officials to manage their administrative and reform efforts, and while it will require resources and will necessarily take time and energy, in the long run the investment will improve the election administration process in the United States.

Making This Work

The world of elections is highly decentralized, and has been since the nation's founding. This tradition of decentralization trumped the desire for strict federal regulation when the Help America Vote Act was being formulated and debated in 2001 and 2002. The lack of regulatory power in the EAC and the lack of mandatory voting system standards highlight this desire to keep the election arena free of strong federal oversight. The lack of a regulatory culture makes changing the dynamic of election quality difficult to achieve. So how can we get there?

One option is to use an incentive approach with states. States and local election officials could be offered resources and technical support—by the EAC or by a third-party entity funded by corporate or philanthropic organizations—on how to implement best practices and better management approaches associated with running elections. In either case, they would be creating a form of a Center for Election Integrity (CEI). The CEI would be offering states and localities what can almost be thought of as an "effective election management in a box" model. The benefit would be that states or localities could be certified as using best election practices, which could qualify them for special funding for elections from various sources.

At the state level, this would involve helping the state develop effective mechanisms for regulating the voting technology used in the state. The CEI would help the state develop an effective quality management model for regulating the acquisition and use of voting technologies in the state and for conducting ongoing analyses of the effectiveness of all voting technologies implemented in the state. The state could also develop processes and procedures for monitoring the quality of elections as implemented by local election officials, since many of the problems associated with the implementation of any voting technology arise because of ineffective management at the local level.

The CEI could also work with election officials to help them improve their management at the local level. This would include identifying the key processes in elections and conducting a threat-risk assessment for these processes, identifying mitigation strategies and security improvements (both physical and process-oriented improvements), and developing appropriate training to assist both management and poll workers in understanding the new system. Again, this process would be designed to improve the management processes that often fail in problematic elections. It would also help local election officials see where the weaknesses are for each process and each voting system that they utilize, because many counties utilize two systems (one for absentee voting and one for precinct-based voting). The CEI could also assist election officials with their customer research and service strategies and help evaluate the extent to which election officials are meeting quality control goals in their administrative and service effort for voters and other stakeholders.

The issues that we see in the election process today are in many ways rooted in the political and historical context of the American experience. The strong role that political parties play in elections dates back to the time in the 1800s when elections were implemented by the political parties, who printed the official ballots, selected the poll workers, and counted the ballots. The voter registration systems and identification requirements of today are rooted in concerns about voting among immigrants at the turn of the past century. Some of the arguments and attitudes about federal regulations of elections date back to the post-Reconstruction period, when federal regulation of elections was associated with the enfranchisement of black voters in the South.

Changes to election processes and procedures also have to be approved by the people who are most affected by election reform: the elected representatives and government executives at the state or federal level. Although members of legislatures are motivated to make good public policy, they are also strongly motivated to keep their jobs. Election reforms, for better or worse, are typically viewed through the lens of politics. One key question, therefore, is whether improvements in election quality and the debate about election reform can be expressed in the language of good policy or whether the current partisan framing of these issues by the media and interest groups will continue.

Chapter 9

CONCLUSION

In this book, we have identified the key issues associated with the use of electronic voting and voting technologies. In the United States, these issues will not disappear; the debates over electronic voting are likely to become more divisive over the next several years, not less. However, the nation has made tremendous progress over the past seven years in understanding many of the issues associated with election technology. Most beneficially, we are starting to understand that voting technology is not the beginning and ending of elections. Because the human factor is a critical part of the elections process, a focus on human–voting machine interactions specifically, and human–voting system interactions more generally, is needed to make elections work. Studies of poll workers and studies of voter satisfaction in the electoral process are being conducted that allow us, for the first time, to understand what makes a voter confident in the electoral process.

We view this book as a beginning, not an ending, in understanding the political, technological, and administrative aspects of electronic voting. Such technologies are only now being used broadly, and the debate over paper trails is changing the landscape in which voting machines are used. The international trials of electronic and Internet voting are expanding our understanding of the technological boundaries associated with democracy and elections. We are seeing changes among the vendors who are in the market, changes in the products they are offering, and even changes in how many of the vendors do business. Data are also only now becoming available on the costs associated with various voting technologies, and such information will help to improve the policy discussions associated with elections. In short, our understanding of electronic voting and voting generally is likely to be different in ten years because of the work of the scholarly community, following the lead of innovative election officials who work hard every day to improve the voting experience of their clients, the American voters.

Our argument regarding electronic voting is easily summarized. We are scientists, and as scientists our lives involve hypotheses and data. Hypotheses are tested, most rejected, and from the process of hypothesis testing we learn how the phenomenon under examination works. For too long, applications of the precautionary principle have held sway in debates about election reform; this has been true in historic debates about suffrage, voter registration, absentee voting, and just about every effort that has been initiated to change the electoral process. The debates about how Americans cast ballots since the 2000 presidential election, in particular the debate about the use of electronic voting systems in polling places, has fallen into this trap. Instead of the precautionary principle, we need a rational, science-based effort to evaluate all voting systems, testing their relative merits, so as to determine their strengths and weaknesses and to improve these voting systems (thus improving the act of voting for Americans). A new science of elections and election administration is needed to form the foundation from which improvements to the American electoral process can begin.

Based on our research about electronic voting, we have ten recommendations that we think will help to initiate this new scientific initiative and thus advance our understanding of the issues associated with the technology itself and how these issues are discussed and debated in the American electoral and policy processes. These recommendations are intended to move the debate over voting technology forward and provide a rational, fact-based environment that will promote better policy making and a more informed debate on election reform.

1. *We need to take research seriously—and fund it seriously.* For as much time and energy that has been spent debating voting technologies since 2000, there has been relatively scant federal funding of research on election administration. The EAC has facilitated some small, but highly idiosyncratic, studies in an attempt to fulfill its legal requirements for studies required by Congress under the Help America Vote Act. However, the broader research agenda on usability, human-machine interactions, security, and support for ongoing data collection of election administration has been supported primarily by limited private—not public—funding. The Carnegie Corporation of New York, the John S. and James L. Knight Foundation, and other foundations have been the primary drivers funding election administration studies. Recently, the National Science Foundation has funded some studies—the most prominent is the ACCURATE project, but the most prolific (to date) federally funded project in terms of published scientific research is the University of Maryland consortium studying human–voting machine interactions—yet the funding allowed for basic research under HAVA has not been appropriated nor, of course, allocated.

We need a broad and comprehensive research and development commitment, with both private and public funding, to sustain real scientific research and to build the framework for a new multidisciplinary and multimethodology study of election administration and voting technology. This research effort has to provide the funding for basic research on a wide array of technological and administrative issues, but also must provide for the translation of basic research into practical procedural and technological solutions to real problems. We need the focus (though not necessarily the financial commitment) that was seen in major applied science projects in our nation's history (like the Apollo or Human Genome projects)—a lasting and dedicated attempt to research and develop better ways for elections to be run in the United States. If, through a combination of private and public funding, we were able to count on just a small percentage of the dollars annually spent on elections in the United States, perhaps a research and development effort of $25 million annually, we suspect that many of the procedural, political, and technological problems that we now see in American elections could be improved or resolved in the next decade.

This investment, although focused on elections, would also likely yield side benefits to the economy much like those related to the Apollo and Human Genome efforts. The research on usability and security would be of benefit not only to the election community but also to firms that engage in secure transactions, person-to-person transactions, and human-machine transactions. For example, there are many transactions that require authentication, auditability, and high levels of public confidence. The research conducted on voter verification in elections will benefit firms that engage in similar transactions. Likewise, anyone who has ever used a credit card or ATM payment system knows that there are dozens of different formats for these systems and the lack of uniformity can be quite annoying. Research on human-machine interactions in elections could lead to improvement in similar transactions, like the payment system process.

2. *We need more systematic data collection.* The study of elections presents us with a contradiction; we are awash in data, but yet also live in a desert of usable data. Most localities collect basic data on elections, although some states fail to collect basic information such as the number of voters who cast ballots in an election. Even when collected by jurisdictions, these data are often not reported at all or are provided in difficult-to-use formats. The 2004 EAC Election Day Survey illustrated this problem. The study had a relatively high nonresponse rate on many questions—some jurisdictions failed to respond to any of the survey questions—and even the questions with a high response rate did not report data uniformly. It is difficult to study elections if basic information, such as the number of ballots in a district that go uncounted, cannot be obtained or calculated.

This problem can be addressed by having a requirement, supported by federal funding, for systematic data collection after every federal election and a uniform standard for data elements that are collected. We have given several talks and written a report for the IBM Center for the

Business of Government on the need for uniform data exchange standards (see Alvarez and Hall 2005b). Uniform standards would ensure that all elements in an election, such as what constitutes an early or absentee ballot, are consistent across states. Uniform standards would also provide that, regardless of the vendor a state or locality used for election administration, the data that were produced would be uniform in format. The federal government should also provide some support to states and localities to offset the cost of this data collection. Given that the states are providing a service to the federal government—running elections that elect federal officers—the federal government should compensate the states for this service and, in exchange, require the production of basic data on elections.

3. *We need to appreciate that all elections have a black-box problem.* One of the interesting things about the debate over electronic voting is that the critics of electronic voting fall into two categories, those who criticize electronic voting but propose paper trails or optical scan balloting as a solution, and those who take a more principled—albeit more radical—position by arguing that all voting systems that utilize secret ballots have the black-box problem. For these critics, paper trails and paper ballots merely provide a false sense of security; we cannot verify that the paper is authentic any more than we can verify the authenticity of the results that come from electronic systems. As a consequence, we should move away from secret ballots and back to the public voting we did at the nation's founding.

As we have noted, the second category of critics has a strong point regarding the problem, even if we do not support their policy recommendation for addressing the problem. All voting systems have an auditability problem that arises from the secret ballot. The secret ballot is explicitly intended to make elections private, by making it impossible to trace a specific ballot back to a specific voter. This makes voting fundamentally different from a transaction such as banking; there is no way to relink voters with their ballot, the voting equivalent to their banking account. Instead, we can audit the election only at a macrolevel; the number of ballots counted equals the number of voters who voted and the ballots counted were secured throughout the chain of custody from beginning to end. The policy rationale for this lack of auditability is simple; we do not want election officials or others in power to be able to coerce, punish, or pay voters based on how they vote. Voters should be able to cast ballots with a clear conscience, without concern that there will be retribution for their actions. However, the result of this policy of voting by secret ballot is that we have to trust the chain of custody procedures used to protect ballots from the point the ballot is printed or programmed through the process of tabulating, reporting, and auditing the election.

We should not be under the impression that some voting technologies are magically not subject to the auditability problem solely because of the medium on which the ballot is cast. Unfortunately, many state statutes for postelection audits that have been adopted in the past six years target only electronic voting, not all voting systems. This creates an unequal policy environment that fundamentally ignores the black-box problems associated with voting on paper ballots. Likewise, voter-verification requirements for electronic voting provide those voters with a different level of service compared to those who vote using optical scan ballots. Voters who cast an optical scan ballot cannot know if the optical scan reader has tabulated their vote accurately; there is not independent verification of the ballot. Because paper ballots have a two-stage process for interpreting voter intent, giving voters verification of how the machine is reading the ballot is just as critical as providing a voter using electronic voting with a paper trail. How voter verification would be provided from a tabulator is an engineering question, but such systems might produce a paper trail or visual verification to the voter.

4. *All voting systems should be held to the same standards for usability, security, and auditability.* Critics of electronic voting have conducted various studies to show how the security of a DRE can be subverted. However, it is odd that fewer studies have done the same with paper ballots or electronic tabulation software used to count paper ballots. After all, paper ballots have historically been the subject of manipulation, if only because paper ballots have been used for more than 200 years in the United States. Even today, one of the most frequently asserted threats to election security is generally absentee voting, which is a paper-based voting system. However, standards for voting system testing and security rarely consider examining the range of threats associated with paper ballots and whether the policies and procedures for chains of custody adequately address the threats. In addition, as we noted previously, voter-verification requirements rarely consider the fact that voters who cast a vote on a paper ballot that is then precinct-tabulated cannot verify that the tabulator counted their vote accurately. DRE voters can look at the paper trail and compare it to the electronic ballot. Paper ballot voters have to hope that the machine tabulated their vote accurately—they receive no feedback regarding how the tabulator counted their ballot.

The response to this, of course, is that the voters' ballots can be recounted if a problem arises. However, the voters have no idea if their ballot—as they cast it—is the ballot that is counted. The practice of remarking ballots—where election officials "correct" ballots that cannot be easily interpreted by a voting machine—is rarely discussed and has never been systematically studied. We have also noted the problems associated with poll workers discerning "voter intent" from the marks on a paper ballot; such problems have rarely been subject to systematic study, even though such interpretations affect the security, auditability, and integrity of elections in states with a voter intent standard. Systematic study of both paper-based and electronic voting would help provide election administrators with data that could be used to compare the risks and threats associated with both methods of voting, as well as methods of mitigating against those risks. Likewise, data on usability, such as those produced by the University of Maryland, that compare electronic and paper-based voting systems can produce important and interesting information that can help election officials evaluate the strengths and weaknesses across voting systems types and across manufacturers within a given class of voting system.

5. *Even though voting systems should be subject to similar requirements, we should recognize that all DREs or optical scan systems are not the same.* It is common to refer to problems with electronic voting without noting that all DRE systems—or for that matter, all optical scan systems—are not the same. As we noted before, studies of voting system security have found differences among electronic voting systems from different manufacturers. In addition, research conducted at the University of Maryland, in conjunction with the University of Michigan and the University of Rochester, has found that voters have very different experiences when comparing different electronic systems or electronic and paper-based systems. The results of their study of write-in voting discussed earlier are an interesting case in point. Voters had very different experiences across voting systems attempting to write-in a vote in a way that would count under state law. Studies of optical scan voting have found that filling in a bubble on a ballot is different for the voter compared to connecting two arrows; the latter method of optical scan voting produces a significantly higher residual vote rate.

If we consider only the primary electronic voting systems that are on the market today, they vary in many ways. Although electronic voting systems by Diebold, ES&S, Hart InterCivic, and Sequoia all record ballots electronically, they have widely varying methods for how voters initiate their voting experience with each company's device, how they input their ballot choices, how those choices are confirmed and verified by the voter, how the choices are stored inside the devices, and how the device communicates the stored ballots at the close of elections. These differences in electronic voting systems mean that it is difficult to make generalized statements about system security, usability, auditability, and so on. As we move forward with the collection of better and higher-quality data, as well as the development of better scientific approaches for studying voting technology, these subtle and not-so-subtle differences in voting systems should no longer be glossed over but should themselves become important variables in voting system studies. General statements like "electronic voting is not secure" should be viewed critically, and we should demand more precise statements about exactly which type of voting system, and what aspects of that voting system, a critic is discussing. Likewise, state laws and regulations—as well as local election official policies and procedures—should differentiate among various voting system manufacturers in the requirements they have for security. The steps and processes for securing a Diebold DRE may be very different than the steps and processes needed to secure a Hart InterCivic DRE, even though both are electronic voting machines. States should have better and more clearly articulated rules for securing specific technologies.

6. *Risk management and mitigation is the cornerstone of an effective election.* In all elections, there are risks that have to be addressed and managed. Prior to the advent of secret ballots, the risks were largely associated with vote buying and voter intimidation. Today, the risks that have to be managed are different. Given the black-box quality of all elections, risks associated with the voting process that have to be addressed are related to ensuring the chain of custody of ballots and the authenticity of the ballots throughout the process. Risk management requires that local election officials understand how the voting technology that they are using interacts with the processes and procedures that they use to manage the election. In many cases, state election laws and election procedures, and the risks they are intended to address, date back to a different era and a different technology, when the state or locality used lever machines or hand-counted paper ballots.

Developing strong procedures and chain of custody rules requires states and localities to understand their processes and the threats to those processes so they can address these threats. This assessment of threats and risks cannot just be a one-shot activity; election officials have to update their efforts constantly as the environment changes. Most importantly, election officials have to train the poll workers and other election workers who actually implement the election to follow the processes and procedures that secure the election and maintain the integrity of all aspects of the electoral process. Many problems that occur in elections with voting technologies have nothing to do with the technology itself but center on the interactions between the poll workers and the technology or the ancillary processes that concern the technology.

7. *Technologies and computing are likely to change dramatically over the next decade.* Policy makers should be careful not to make policy decisions that inhibit innovation and create legal "technology locks." When Congress passed the original Clean Air Act, the law required reductions in polluting emissions from automobiles. However, the law

did not state how the reduction was to be achieved. This challenge to reduce emissions fostered competition and innovation, resulting in the invention of the catalytic converter. Had Congress established in policy how the reduction in automobile emissions was to be achieved, the resulting solution, which was both elegant and cost-effective, would likely never have been developed because the policy decision about how the problem was to be addressed would have already have been made.

The current debate over election auditability creates a similar challenge. The goal is to create a system that can instill confidence among voters and create enhanced auditability. We have concerns about laws requiring voter-verified paper audit trails because these laws stymie innovation into other solutions to the security and auditability problem. We are already seeing that paper audit trail laws are no panacea. Paper is hard to count, jam-proof printers difficult to engineer, and chain-of-custody procedures for these systems can be difficult to implement. But these problems do not lead us to assert that voting systems should not have paper trails. As an interim solution, paper trails may be a mechanism for creating independently auditable voting for electronic voting machines. Yet we should remember that the goal of the process is not to require a paper audit trail but to create an election system where voters can verify their choices and where those choices can be independently audited. Even in the short term, the best way to achieve auditability may be through a cryptographic verification system, using the technologies that have been developed by companies both in the United States and internationally.

Interestingly, this is another area where having uniform electronic data transaction standards would enhance the electoral process. Uniform data standards would allow third-party companies to develop "plug-and-play" components for election auditing that would work with any voting system. The output of the voting machines—the votes for specific candidates—would be rendered in the same output format regardless of the manufacturer. This output could then flow into any vote auditing or verification system that was designed by an entrepreneurial company. In fact, one principle of security is the two-man rule; having an auditing system that comes from a competitor of the voting system manufacturer helps to ensure that the coordination needed to "hack" the system is more difficult to engineer.

8. *We need to study the human factor in elections.* In the 2006 election, one of the biggest problems related to electronic voting occurred in the Thirteenth Congressional District race, in Sarasota County. The very high undervote rate that occurred in that congressional race was largely attributed to a ballot design problem in Sarasota County's electronic voting machines. The "banner" on the electronic ballot, designed to inform voters that the ballot was shifting from federal races to state races, appears

to have drawn voters' attentions away from the race above the banner and voters subsequently skipped that race on the ballot. The banner effect as an issue in ballot design is an example of a human factors problem in elections. Individuals interact with technologies and processes in ways that we might not expect. The butterfly ballot is an analogous case to what happened in 2006 with the banner on the electronic voting machine. The voters in Palm Beach County, Florida, interacted with the ballot in a specific way that affected the results of the election.

We are only now beginning to study human factors in elections: voter–machine interactions, poll worker–machine interactions, and voter–poll worker interactions. Work by the University of Maryland consortium, and by MIT's Ted Selker and his research group, has identified a variety of issues associated with human-machine interactions. Conducting write-in voting is easier on some technologies and some electronic voting machines than others. Ballots with a straight-party voting option have a different set of issues associated with them compared to ballots without a straight-party voting option. The ability of voters to use different verification systems is also variable; not all vote-verification systems are equal from a human factors standpoint. Work conducted at the University of Utah and Brigham Young University has examined voter–poll worker interactions and found that the quality of this interaction affects the voter's confidence in the fairness of the electoral process. Other social scientists are studying usability issues related to ballot order effects, ballot completion effects across voting systems, and proximity effect in multicandidate races, each of which affects the voters' ability to cast a vote as they intended. In short, we know these interactions are important; designing systems that address the human factor needs is critical for improving elections.

Unfortunately, human factors in elections are also often "locked in" because of legal requirements. States may have specific rules for the font size or formatting of ballots that were designed for a paper system or lever machines but are no longer appropriate for today's elections or today's voting technologies. New York State is an excellent case in point; the "full-face ballot" requirement in the state would seem to be specifically designed for lever machines. There is little rationale for using such full-face systems today. In fact, there is some research suggesting that full-face ballots are discriminatory against some voters.

9. *We need a modern regulatory scheme that reflects the realities of today's elections.* As we are writing this conclusion, one of the independent testing authorities that certify voting systems has lost its accreditation from the EAC to certify voting systems. The current testing process is not designed to meet the challenges we have today with voting systems. The current regulatory scheme is applying a horse-and-buggy model to a

dynamic and rapidly evolving technology—and every day, this disjunction between the current regulatory scheme and the realities of the technology seems to be widening.

For starters, the current regulatory scheme is voluntary, slow to evolve, and lacks transparency. Each of these problems should be addressed before the next iteration of the voting systems standards process is initiated. Regulation of how voting systems are tested and certified for use must become an open and transparent process, though conducted in a way that preserves any proprietary or critical intellectual property of a voting system vendor. But information on how systems are tested, and how the systems fare in the testing, must be made available to the public in a timely and useable manner. The regulatory scheme must become dynamic, and must allow for continual testing, certification, retesting, and recertification. When bugs or problems are found in any voting technology (ranging from the technologies used for behind-the-scenes election administration to the systems used by voters to cast their ballots), we have to allow election officials and vendors to patch those bugs promptly, to have those patches tested quickly and reliably, and to allow for the dissemination of information about both the problems and the solutions so that all stakeholders can be confident that the machinery of the American electoral process is working as best as we can engineer.

A case in point for the unwieldiness of the current system testing and certification process is the ongoing debate over Diebold's voting system. Consider, for a moment, what would happen if Diebold was to announce each time its critics found a problem that it had addressed the problem in its software. Each time, Diebold not only would have to go through a long federal certification process but would also have to have its software certified by the various states that have state-specific certification requirements. There would be no way for Diebold to have its software certified easily and in the hands of its customers in time for the election. Unfortunately, the certification process creates an adversarial process where vendors defend products because there is not an easy way of addressing the problems within the existing certification process.

Finally, we have to move away from voluntary standards, certification, and testing. All election technology that is used for federal elections must meet minimal federal standards for accessibility, accuracy, auditability, interoperability, security, reliability, usability, and verifiability. Of course, if states wish to go further than the required federal guidelines, we applaud that effort—because it is likely to lead most of the voting system vendors to develop products that will meet the strictest state standards, especially if the strictest standards are in large states or are adopted by groups of states.

10. *We need to build collaborations*. When we first started down the path that lead to this book—a path that dates back to before the 2000 presidential election and spans more then eight years—it was clear then that figuring out how new and evolving technologies can be used in our electoral process, as well as identifying the human factors and training associated with implementing these technologies, required new paradigms and new types of collaborations. There is no existing academic field that studies election administration or voting technology, but those of us who are now working in these areas find that academic research and teaching require collaborations between social scientists, engineers, computer scientists, lawyers, historians, practitioners, and many others in the academy. These working relationships are difficult to build and maintain in the structure of any university, not to mention across universities. Within universities, different departments and units have different incentive structures that hinder effective collaboration. Across universities, collaborations can be hindered by things as simple as budgeting and accounting rules that govern sharing of research funding. Moreover, the work of studying election administration is seen by many in academia as too applied and atheoretical. Even within political science, there are difficulties in getting various parts of the discipline to take election administration seriously. For individuals who study public administration, election administration is viewed as "electoral studies," but scholars of electoral studies view election administration as "public administration"!

Furthermore, many election officials perceive academic researchers as self-interested, research-grant-motivated, and poorly informed—meaning that it is often very difficult to develop collaborations between the research and election official communities, even though those collaborations are critical for both academics and election officials. A third layer of complexity arises from strained relationships between the voting system industry and the research community. One well-publicized bad experience between an election official and a researcher, or between a vendor and a researcher, can make it difficult for other scholars to work with vendors and practitioners, who become understandably reluctant to cooperate with scholars who do not appreciate the complexities of the election process.

Building collaborations requires building trust, and the latter is sorely lacking in the heated debates about election reform and voting technologies. Trying to develop the scientific and policy agenda that will lead to lasting and effective solutions for the administrative, political, and technological problems that exist in the American electoral process will require working across traditional academic and professional boundaries. Efforts to bring the key research and policy stakeholders together—like the periodic face-to-face events sponsored by the Caltech/MIT VTP, the National Academies of Science, or the American Association for the Advancement of Science—should continue. But others need to get involved in the development of collaborations, including the private foundations, the Election Assistance Commission, and the voting system industry. Institutionalizing collaboration will prove key for lasting election reform in the United States.

We don't pretend to think that we have all of the answers, or that these ten recommendations will provide a magic bullet that will fix all of the problems with voting technologies and election administration in the United States overnight. Instead, these recommendations should help to improve the science, administration, and practice of American elections, thereby helping to improve the integrity of our brand of democracy, and also helping to ensure that the confidence of all stakeholders is as strong as possible. We do hope that this book stimulates a thoughtful, fact-based, rational discussion of the full arrays of issues associated with election administration and voting technology in the United States. Given the amount of heat that has been generated by harsh rhetoric and accusations over the past seven years, a calm debate that leads to better elections would be a pleasant change.

NOTES

1. Bousquet 2004, http://www.sptimes.com/2004/07/29/news_pf/State/GOP_flier_questions_n.shtml.

2. http://www.npr.org/templates/story/story.php?storyId=4131522.

3. Unfortunately, there is very little published research on aspects of election administration, like absentee ballot return rates, so it is impossible to produce a reliable national percentage of absentee ballots counted of those returned. The rates at which absentee ballots are counted appears to vary considerably across election jurisdictions, even within the same state: for example, according to data from California counties compiled by the U.S. Election Assistance Commission in its 2004 "Election Day Survey," the rate of uncounted absentee ballots ranges from a few percent in many counties, to double-digit non-counted rates in places like Riverside County (11.1%), Sonoma County (11.8%), San Joaquin County (13.4%), Tulare County (17.4%), and Yuba County (27.4%); see http://www.eac.gov/election_survey_2004/statedata/California_Jurisdictions.html for these data.

4. See http://www.collinscenter.org/initiatives/initiatives_show.htm?doc_id=105009.

5. For analysis of Georgia's transition to e-voting in 2002 and the immediate effect on residual vote rates, see Stewart (2004). Georgia's secretary of state, Cathy Cox, released data on 18 November 2004 depicting a drop in the presidential residual vote rate (or the percentage of ballots that showed no choice in the presidential race) from 3.5 percent in 2000 to 0.39 percent in 2004. The press release issued by Cox's office can be found at http://www.sos.state.ga.us/pressrel/111804.htm; there interested readers can download data from the 2000 and 2004 elections (2004_pres_undervote_analysis.xls).

6. *Tampa Tribune*, 2004, 18. A parallel debate that we suspect will become much more involved in the near future concerns electronic statewide voter registration systems. The Help America Vote Act of 2002 requires that states move to centralized and computerized voter registration lists, but that same piece of legislation is silent as to the specifics of how these systems are to be developed, what standards they should meet, and how they might be tested and certified for use, raising concerns about potential problems with these new electronic voter registries (Alvarez 2005b).

7. There are also voting technologies that blur the distinction between punch card and optical scan ballots; an example is the "InkaVote" system used in Los Angeles County, California. There the voter has a ballot card that looks virtually identical to a punch card ballot, and the voter uses the same type of vote recorder device in the polling location to help identify the correct spots on the ballot to mark her vote preference, but instead of punching a hole in the ballot

the voter makes a mark with an ink pen and the ballot is then tabulated later using an optical scan device.

8. We group stand-alone and LAN electronic voting together as there are some electronic voting solutions (for example, the Hart InterCivic eSlate voting system) that involve the use of a network within the polling place, as the voting devices used by the voters are all daisy-chained to a central serverlike workstation controlled by election officials. Many other electronic voting devices are stand-alone, in the sense that the voting machines used by voters are not collected to any central workstation in the polling place.

NOTES TO CHAPTER 2

1. Letter to the Editor, *Los Angeles Times*, written by Catherine Getches (2004).

2. Readers who are interested in full book-length discussion of the history of voting technology in the United States should consult Saltman (2006).

3. For example, many contemporary observers lament the long ballots that many voters face in each election, arguing that such long ballots can produce voter fatigue and apathy. It is interesting to note that these same arguments were made at other points in American political history. Ludington's essay in the 1911 *American Political Science Review* describes the "Short Ballot" movement, "which aims to simplify State and local government by a reduction in the number of elective officers, has been widely recognized and discussed in all parts of the county, and has secured powerful support in a number of states" (79). There is a debate in the research literature on the initiative process that has cast doubt on the claim that long ballots necessarily lead to voter fatigue and reductions in participation (see Magleby 1984 or Tolbert, Grummel, and Smith 2001). Also, while there is no doubt that modern presidential campaigns have been quite negative in recent memory, it is also the case that negativity is not a new feature of American political campaigns: a notorious example was the presidential election of 1884, when Republicans attacked Democrat Grover Cleveland for fathering an illegitimate child (a claim which he acknowledged) and Democrats attacked Republican Blaine on claims of his involvement in various corruption schemes. See Ansolabehere and Iyengar 1995 and Patterson 2003, for arguments about the relationship between the negativity of modern campaign politics, voter apathy, and turnout.

4. Bensel (2004) notes that voice voting ("vica voca" voting) still existed in a handful of states through the Civil War period and its immediate aftermath; the last state to use voice voting was Kentucky, which ended the practice in 1891. For further discussion, see Bensel 2004, 54–57.

5. For examples of ballots from this period in California, see the presentation developed by Melanie Goodrich, "19th Century Ballots from California," http://vote.caltech.edu/ballots/huntington, and her research study, Goodrich 2004.

6. For example, it is not uncommon to today find historical partisan paper ballots from this period that have small pieces of paper glued over the names of certain candidates, either obscuring the underlying candidate's name or replacing

the underlying candidate's name with another candidate. An example of such a ballot from 1859 that was studied by Goodrich (2004) can be viewed at http://vote.caltech.edu/ballots/huntington/1859-6and7.htm.

7. Much has been written about fraud and paper ballots. For a recent examination of the use of paper ballots in the early decades of American history, see Bensel 2004.

8. This process is described in Keyssar 2000, 28.

9. For an example of a lever machine's presentation of a modern ballot see http://vote.caltech.edu/media/documents/brooklyn5.jpg, which is a photograph of a lever machine ballot from Brooklyn, New York, during the 2004 general election. Note the presentation of this ballot. A voter can either pick specific candidates for each race by tripping levers across rows of candidates or vote a straight party ballot by selecting an entire column of candidates.

10. With today's digital technologies, it could be relatively easy for a voters to record their selections within a voting booth digitally, and also to record the casting of the vote digitally—and to email the digital recording to an accomplice. For this reason, many places do not allow cameras, cell phones, or other digital recording devices into polling places, but it is difficult for busy polling place workers to enforce these regulations easily.

11. Susan King Roth (1998) examined the usability of these machines in depth and found that being short or having physical dexterity issues resulted in many voting errors.

12. There is an array of other problems with mechanical lever voting machines, especially administrative problems: they do not allow for an audit trail, or a real vote recount, they are difficult to use in situations where write-in ballots are allowed, and they are typically very large and hence pose transportation and storage problems. Saltman (1988) discusses these administrative problems in detail.

13. A copy of one such "Votomatic" punch card ballot, from Franklin County, Ohio, can be seen at http://www.yourvotecountsohio.org/images/ballot_front.jpg.

14. For an example of this type of punch card ballot, see http://vote.caltech.edu/media/documents/ballotfront.pdf and http://vote.caltech.edu/media/balloback.pdf.

15. There is a hybrid punch card, optical scan voting system that has been adopted for use in Los Angeles County, California, called InkaVote. This device uses a card that is identical in size and format to a traditional punch card, but instead of indicating their vote by way of a punch through the card, voters use an ink pen to mark a small dot on the punch card. The election officials can use much of the same equipment as they would use with a punch card voting system, including the vote recorder devices used in polling places and other equipment. Some (Brady 2004) have criticized the accuracy of the InkaVote system, and there have been other administrative problems associated with its use in a recent Los Angeles City election, where election officials from the city clerk's office (the city uses the county's voting machines when the city runs an independent election) "overmarked" a large number of ballots (McGreevy, 2005, http://www.latimes.com/news/politics/la-me-vote22mar22,1,6446745.story?coll=la-home-headlines.

16. Some attention was focused on this issue as the result of problems with SAT test scores hat were widely publicized in 2006 (http://electionupdates.caltech.edu/

2006/03/does-wet-paper-pose-threat-to-optical.html). Some who have studied optical scan ballots have raised concerns about humidity's effects on the ballots, and there have been reports in the media of how high humidity levels may have affected the performance of optical scan voting systems (http://electionupdates. caltech.edu/2006_03_12_archive.html).

17. Herron and Lewis 2006, http://www.dartmouth.edu/~herron/Pasco.pdf.

18. Our preliminary look at the initial results from Sarasota County is available at http://electionupdates.caltech.edu/2006/11/whats-going-on-in-sarasota-countys.html.

19. Ongoing research by Matthew Doig and Maurice Tamman, staff writers for the *Herald Tribune*, concludes that ballot design is the most likely explanation for the high undervote in Sarasota County's thirteenth Congressional District race (http://www.heraldtribune.com/apps/pbcs.dll/article?AID=/20061205/NEWS/ 612050604/-1/xml). The Doig-Tamman analysis is supported by research from Frisina et al. 2007. However, expert witness reports filed by Charles Stewart III and Dan Wallach, in support of litigation contesting the outcome of the race, argue that other factors might have caused the high undervote by electronic voters in this situation (http://electionupdates.caltech.edu/2006/11/sarasota-cd-13-court-filings.html).

20. See Alvarez, Sinclair, and Hasen 2006 and Ho and Imai 2004 for debate over the practicalities of randomizing the ballot for every voter.

21. This work can be found at http://www.capc.umd.edu/rpts/VotingTech _par.html.

NOTES TO CHAPTER 3

1. See Beck 1992, 1998, 1999; Mythen 2004; Pidgeon, Kasperson, and Slovic 2003; Slovic 2000 for a complete discussion of the risk society and the social aspects of understanding risk.

2. For an examination of the business implications of this decision, see http://www.usatoday.com/money/industries/health/drugs/2005-01-13-drug-sales _x.htm.

3. See, for example, data from the World Organisation for Animal Health, http://www.oie.int/eng/info/en_esbincidence.htm, and from the Centers for Disease Control and Prevention, http://www.cdc.gov/ncidod/diseases/cjd/bse_cjd.htm.

4. The problem of BSE in Europe and the United Kingdom is covered in numerous studies of the risk society, including Beck 1998; P. Harris and O'Shaughnesy 1997; Mythen 2004.

5. Giddens 1998, taken from Mythen 2004.

6. "Food-Related Illness and Death in the United States," http://www.cdc.gov/ ncidod/eid/vol5no5/mead.htm.

7. See Karatnycky 2005 for a detailed accounting of the events of November and December 2004 in the Ukraine. The allegations of fraud primarily involved absentee voting fraud and manipulation of ballot tabulation. Karatnycky wrote "Numerous reports indicated that roving teams of voters, tens of thousands in all, were being transported in trains and buses from polling station to polling

station, each armed with multiple absentee ballots" (36). He also wrote of evidence of "late-night manipulation of data in the CEC's [Central Election Commission] computer server."

8. See Hall 2003; testimony before the Senate Judiciary Committee, http:// judiciary.senate.gov/hearing.cfm?id=909; and Fortier and Ornstein 2004.

9. The Caltech/MIT Voting Technology Project issued preliminary studies on the reliability of voting technologies in March and then February 2001, and a final report in July 2001. All are available from http://www.vote.caltech.edu/reports.

10. Before 2001 there was little peer-reviewed research on electronic voting; much of the published research focused on remote Internet voting. For writing in the area of electronic elections, see Neumann 1990; Shamos 1993; and Mercuri 2000.

11. The Administration and Cost of Elections (ACE) Project – a collaboration of IFES, an international nonprofit organization that supports the building of democratic societies; the International Institute for Democracy and Electoral Assistance (IDEA); and the United Nations Department of Economic and Social Affairs (UNDESA)—has identified eight principles for effective vote counting. See http://www.aceproject.org/main/english/vc/vc20.htm.

12. The association of the black-box problem with e-voting has been popularized by B. Harris (2004), and the associated Web site, http://www.blackboxvoting.org.

13. For example, Democratic House member Brian Baird wrote in the *Washington Post* (27 November 2004, A31) about recent instances of last-minute language being inserted into legislation in the Congress. Baird wrote that members often have "only a few hours to read bills that are thousands of pages in length and spend hundreds of billions of the people's dollars."

14. There are cases where a ballot can be traced to a voter. In the state of North Carolina, a voter's absentee ballot can be traced back to the voter using a ballot stub number that is on the ballot and recorded on the absentee ballot request. This rule is designed to thwart absentee voter fraud. Also, voters can make their ballot traceable in some cases by casting a write-in ballot. The practicality of this, however, can be modest because in many jurisdictions (e.g., Utah) the votes of only *registered* write-in candidates are counted in the official tabulation. Finally, the voter may be the only individual in a precinct to vote for a certain third-party candidate.

15. The same problem does not necessarily arise in private elections. One common form of private election is a corporate proxy ballot (which virtually everyone who either directly owns corporate stock or holds shares in a mutual fund has received). Here, the voter's choices are marked on a ballot that has a variety of tracking indicators right on the ballot, thus making it possible at all times to know which ballots have been tabulated and which have not. A similar situation arises in some presidential primary or caucus votes in recent years, for example, the 2004 Michigan Democratic presidential caucus, where the state party allowed votes to be cast online. In this caucus vote, however, our understanding is that voters' intentions were recorded along with their identifying information; hence these votes were cast not in secret but in a situation that makes the process potentially more secure. If a voter was not eligible to cast a ballot (or cast more than one ballot), this could be rectified in the final vote tabulation.

16. Many of these protections on electronic transactions for consumers are in the Fair Credit Reporting Act and the Electronic Fund Transfer Act. For additional information on both these acts, and on federal regulations regarding consumer protections of electronic transactions and commerce, see the Federal Trade Commission's Web site on consumer credit, http://www.ftc.gov/bcp/conline/edcams/credit/rules_acts.htm.

17. Of course, the same basic plot was used in the recent remake that starred Denzel Washington and Meryl Streep, though we prefer the original in this case. The novel that both movies are based upon was written by Richard Condon and published in 1959.

18. This section uses Kohno et al. 2004 as its basis.

19. The story regarding how the code was obtained can be found in the essay, "The Case of the Diebold FTP Site," http://www.cs.uiowa.edu/~jones/voting/dieboldftp.html.

20. "Risk Assessment Report: Diebold AccuVote-TS Voting System and Processes," 2 September 2003, is available at http://www.dbm.maryland.gov/dbm_publishing/public_content/dbm_search/technology/toc_voting_system_report/votingsystemreportfinal.pdf, executive summary, p. III (emphasis in original).

21. This report, "Trusted Agent Report: Diebold AccuVote-TS Voting System," 20 January 2004, can be found at http://www.raba.com/press/TA_Report_AccuVote.pdf.

22. Feldman, Halderman, and Felten 2006, http://itpolicy.princeton.edu/voting.

23. Election Science Institute, "DRE Analysis for May 2006 Primary, Cuyahoga County, Ohio," http://bocc.cuyahogacounty.us/GSC/pdf/esi_cuyahoga_final.pdf. We were part of the research team involved in this study; primarily we focused on the analysis of the precinct incident reports.

24. Pacific Design Engineering, "Sequoia Voting Systems Vulnerability Assessment and Practical Countermeasure Development for Alameda County," 4 October 2006.

25. "OIG FINAL REPORT: Voting Systems Contract RFP 326," http://www.miamidadeig.org/reports/voting%20final%20report.pdf.

NOTES TO CHAPTER 4

1. There is surprisingly little research on optical scan voting systems, and their effects on outcome measures like residual votes. For summaries of the research to date, see Alvarez, Sinclair, and Wilson 2004; Knack and Kropf 2003; Tomz and van Houweling 2003.

2. For an example of this related to the abortion policy domain, see Saletan, 2003.

3. See, for example, Brown 1990; Inglehart 1984; Lanoutte 1990; Rothman and Lichter 1982; and Weart 1988.

4. Baumgartner and Jones 1993, 64–65; taken from Weart 1988.

5. Their work and its implications are well covered in Pidgeon, Kasperson, and Slovic, 2003.

6. These data were collected by Ms. Amy Sullivan in May 2005 at the request of the authors. Given that LexisNexis is a dynamic database, even in its historical files, efforts to replicate these findings may produce slightly different results.

7. We show below in chapter 7, based on survey research that we have done, there is polarization in the attitudes of voters regarding electronic voting.

8. http://www.aclu.org/VotingRights/VotingRights.cfm?ID=7090&c=166.

NOTES TO CHAPTER 5

1. See Alvarez, Hall, and Roberts, forthcoming.

2. http://www.electoralcommission.org.uk/templates/search/document.cfm/6319 (2002, 15).

3. For example, Pippa Norris has presented several papers examining theories of voter turnout using these data on her Web site, http://ksghome.harvard.edu/~pnorris/Articles/Articles%20conference%20papers.htm#Conference%20Papers%20on%20Elections%20and%20Public%20Opinion. Many of the original reports are available from the Electoral Commission Publications Web site, http://www.electoralcommission.org.uk/about-us/researchpub.cfm.

4. U.K. Electoral Commission, "Delivering Democracy? The Future of Postal Voting," London, August 2004.

5. The full array of studies and research conducted by the U.K. Electoral Commission can be found at http://www.electoralcommission.org.uk/.

6. See http://www.electoralcommission.org.uk/media-centre/newsreleasereviews.cfm/news/442

7. A summary of the European experience with Internet voting can be found in the manuscript "E-voting and Electoral Participation" by Alexander Trachsel, European University, Florence, Italy.

8. Georgia continues along this path. As we revise our book for publication the state just conducted pilot tests in a small set of counties to test the utility of the voter-verified paper audit trail.

9. The evaluation report for this trial can be found at http://www.vote.caltech.edu/Links/AlexandriaReport.pdf.

10. The early major pieces of federal legislation dealing with this class of American citizens are the Federal Voting Assistance Act of 1955 and the Overseas Citizens Voting Rights Act of 1975; these were modified by the National Defense Authorization Act for Fiscal year 2005. The Help America Vote Act of 2002 also contains many provisions dealing with UOCAVA citizens.

11. That is, the SERVE system necessitated a process that LEOs could use to make the system (or just components of the system) available for their authorized staff to access and use. This involved an enrollment and authentication process, similar to that used by participating voters, whereby LEO staff would be issued credentials to use the SERVE system or components of the system.

12. A digital signature is simply a form of electronic identification that would have been provided to SERVE users upon completion of the enrollment process. This electronic identification would then be used each time a user engaged in an interaction with the SERVE system.

13. See the Department of Defense Web site, http://www.defense.gov/news/ Feb2004/n02062004_200402063.html, for a summary of this decision.

14. See http://www.servesecurityreport.org.

15. See, for example, chapter 2 of *Point, Click and Vote*, where we presented the arguments of Jefferson and Rubin against Internet voting, critiques that were written well before January 2004.

16. Barstow and Van Natta 2001, These data were also examined in Imai and King 2004.

17. GAO 2001a, http://www.gao.gov/new.items/d01704t.pdf; Elections: Voting Assistance to Military and Overseas Citizens Should Be Improved, GAO-01-1026, 28 September 2001, http://www.gao.gov/new.items/d011026.pdf; Overseas Absentee Ballot Handling in DoD. Department of Defense, Office of the Inspector General, 22 June 2001. http://www.dodig.osd.mil/audit/reports/ fy01/01-145.pdf

18. http://www.fvap.gov/services/survey.html.

19. This special directive was issued by Shelley's office on 12 September 2003. It detailed seven conditions that must have been met for a voter to able to use fax technologies to cast a ballot: the voter had to be a "special absentee voter," the voter needed to waive the right to a secret ballot, the ballot had to be received by the election official before the closing of polls on election day, the election official had to try to protect the secrecy of the faxed ballot, the voter needed to provide all legally required information on the ballot declaration, the voter's eligibility could be established, and the ballot was preserved for postelection auditing purposes.

20. The analogous threats between Internet and absentee voting are presented in chapter 5 of *Point, Click, and Vote*.

21. These arguments also ignored the architectural design of the SERVE system, which from initial plans to final product had system security, integrity, and reliability as central features of the system. Unfortunately, at this time the system details still have yet to be made public, and thus key features of the system design that should have mitigated or minimized many of the potential security risks were not part of the public and media discussion following release of the report in late January 2004.

22. For example, see Bynum 2000.

23. See, for example, *California Democratic Party v. Jones*, 530 U.S. 567 (2000), where the Supreme Court found that California's "blanket primary" violated the First Amendment rights of political parties to free association. There, Justice Scalia, writing for the majority, argued that "Representative democracy in any populous unit of governance is unimaginable without the ability of citizens to band together in promoting among the electorate candidates who espouse their political views. . . . In no area is the political association's right to exclude more important than in the process of selecting its nominee. That process often determines the party's positions on the most significant public policy issues of the day." See also *Eu v. San Francisco County Democratic Central Committee*, 489 U.S. 214 (1989), and *Democratic Party of United States v. Wisconsin ex rel. La Follette*, 450 U.S. 107 (1981).

24. See http://www.gmu.edu/~action/states/mi.html for useful background information on recent elections in Michigan.

25. The Michigan Democratic Party debated whether to schedule its caucuses on 27 January 2004 (the same day as New Hampshire) or on 7 February 2004. The early January date, however, would have violated national Democratic Party rules that preserve the first-in-the-nation status for Iowa and New Hampshire.

26. See Pickler 2003a, http://www.usatoday.com/tech/news/techpolicy/2003-11-20-net-voting-mich_x.htm.

27. See Hoffman 2003, http://www.detnews.com/2003/politics/0311/01/politics-312975.htm and 2003b, (http://www.mercurynews.com/mld/mercurynews/news/politics/7328526.htm?1c).

28. Despite the public declaration of Democratic affiliation, the voter need not be a member of the Michigan Democratic Party to participate in the caucus; Michigan does not require partisan voter registration. As a practical matter, the procedure here is consistent with Michigan's practice of having an open primary, where any registered voters can participate in the Democratic caucus as long as they were willing to state they were Democratic when the cast their ballot. Also, note that this procedure allows for votes to be cast by those who could for various reasons be eligible to register to vote in Michigan by the November general election, but who were not necessarily eligible in February. This provision would have made it possible for individuals who were not eighteen years of age in February, but who would be eighteen by the general election registration deadline, to participate in the Democratic caucus.

29. Michigan Democrats use a proportional allocation procedure. The eighty-three district-level delegates (and fifteen alternatives) that Michigan was allocated for the national convention were distributed across the fifteen congressional districts in the state based on a formula that gave equal weight to the districts' 2000 Democratic presidential vote and their 2002 gubernatorial vote. Delegates were then apportioned to presidential candidates based on the caucus results in their congressional district, using a proportional allocation formula. That formula allocated district-level delegates and alternatives in proportion to the percentage vote won by each candidate in the caucus, except for candidates who received less than 15 percent of the vote. Candidates receiving less than 15 percent of the caucus vote in a district did not receive any district-level delegates or alternatives.

30. See http://www.electionservicescorp.com for details on Election Services Corporation. In 2003 reports surfaced indicating that Election.com, a company developing Internet-based voting systems had sold assets relating to the development of public-sector elections to Accenture (Harrington 2003). The remaining assets of what had been Election.com appear to have become Election Services Corporation.

31. Unless otherwise noted, all of the data in this section of the report on the Michigan caucus comes from the Michigan Democratic Party (http://www.mi-democrats.com).

32. We compiled the number of mail or Internet ballot requests from data made available by the Michigan Democratic Party on the number of ballot requests and the number of Internet ballots, by age of the voter. Media reports give the number of mail or Internet ballot requests as approximately 123,000 (see Weeks 2004, http://www.detnews.com/2004/0402/06/all-56762.html).

33. Alvarez and Hall 2004, 129. There were also 4,174 votes cast in the Arizona primary from Internet voting stations in polling places throughout the state on election day.

34. For example, in a press release dated 1 January 2004 ("Vote Now in MI's Feb. 7th Presidential Caucus"), Mr. Brewer was quoted as saying, "This will be the earliest, easiest to participate in and best-attended Caucus in the history of the Michigan Democratic Party."

35. For example, the party worked with a cab service in Detroit that provided free rides for senior citizens to polling places on election day. More innovative was the party's collaboration with www.publius.org; voters could either go to that Web site, or call a toll-free number, where they could verify their registration status for the caucus, determine their polling place, obtain a sample ballot, get information on the candidates, or find instructions on the process.

36. See http://www.cnn.com/ELECTION/2004/pages/results/states/MI/I/02/epolls.0.html.

37. Avi Rubin made this claim on a radio program where he appeared with one of the authors.

NOTES TO CHAPTER 6

1. This story was told to Hall by Charles Stewart at the "Voting, Vote Capture, and Vote Counting Symposium," Kennedy School of Government, Cambridge, Massachusetts, 1 June 2004.

2. A voter *intentionally* not marking a ballot in a specific race is also a form of choice. It is often a comment by the voter regarding the choices offered, the length of the ballot, or other factors.

3. The issue of confidence and the controversy are discussed in the 2001 reports of the National Commission on Federal Election Reform and the Caltech/MIT Voting Technology Project. Each author worked with one of these groups at the time the reports were produced.

4. See, for example, Alvarez, Sinclair, and Wilson 2004; Brady et al. 2001; Bullock and Hood 2002; Herron and Sekhon 2001; GAO 2001b.

5. Some critics of electronic voting have argued that electronic voting makes us all blind, in that the sighted cannot know if their ballot was counted accurately on this technology because of its black-box aspects. We would note only that no voter—on any voting platform—can know if a ballot was counted accurately unless there is the only one voter in a precinct because any voting technology can have miscounted a ballot or have ballots subjected to fraud.

6. This transition spawned a minor controversy, when in the first major election the new electronic voting machines were used in Georgia, Republican candidates beat Democratic candidates in two prominent top-of-the-ticket races, taking the positions of governor and U.S. senator from Democratic hands. Some observers saw these Republican victories as surprising, at odds with some preelection polls (Gumbel 2005, 237–238). But other analyses, especially Stewart's (2004) statistical analysis of the relative performance of voting systems

in Georgia before and after this transition, indicate that the move to the Diebold electronic voting systems led to "recovery" of potentially lost votes in more rural, low-income and low-educational attainment, and African American counties.

7. It is interesting to note, however, that the residual vote rates for paper-based systems are improved by the remaking of ballots. The actual residual vote rate—as reflected in the actual ballot marking done by voters—is unclear.

8. See the electronic version of the report at http://www.eac.gov/election _survey_2004/toc.htm.

9. The complete ESI report, "DRE Analysis for May 2006 Primary, Cuyahoga County, Ohio" is available at http://bocc/cuyahogacounty.us/GSC/pdf/esi _cuyahoga_final.pdf.

10. The fifth category involved residual cases, 21.6 percent of reported cases, in which there typically was not enough information to determine the precise nature of the reported problem.

11. See the Center for American Politics and Citizenship Web site, http://www .capc.umd.edu/, for these reports.

12. Democracy at Risk: The 2004 Election in Ohio, http://www.democrats. org/a/2005/06/democracy_at_ri.php. See especially "Section VI: Turnout, Residual Votes and Votes in Precincts and Wards," pp. 3–4.

13. These events are chronicled by Traugott, Highton, and Brady 2005.

14. The Great Debate still raged well after the election; see http://www .appliedresearch.us/sf/epdiscrep.htm for updates.

15. See the letter from Walter Mebane at http://macht.arts.cornell.edu/wrm1/ commondreams/commondreams.html.

16. The study was authored by Michael Hout, Laura Mangels, Jennifer Carlson, and Rachel Best of the University of California, Berkeley, and was released on 12 November 2004. However, the study appears to have been removed from the Web site at Berkeley where it was originally posted, with no explanation for removal provided.

17. See Conny B. McCormack, "Voting System Comparisons/Evaluation of Touch Screen Pilot Project/Recommendation for the Future," 1 January 2001, (http://lavote.net/general/vs_and_chad/vs_and_chad.html, McCormack reports on the process leading up to Los Angeles County's pilot test in November 2000 of touch screen voting systems and lays out a transition path toward future countywide implementation of similar systems.

18. See Secretary of State's Ad Hoc Touch Screen Task Force Report, 1 July 2003, http://www.ss.ca.gov/elections/taskforce_report_entire.pdf.

19. Shelley's February directives are at http://www.ss.ca.gov/elections/security %20measures%20for%20touch%20screen%20(dre).pdf.

20. The complete report outlining the results of the state's parallel monitoring can be found at http://www.ss.ca.gov/elections/voting_systems/2006_nov_pmp _findings_final_rpt.pdf.

21. For a brief description of the process in New Jersey, see that state's Technical Services Bureau, of the Division of Gaming Enforcement in the attorney general's office (http://www.state.nj.us/lps/ge/tsb_enforce.htm). In Nevada, the similar testing and inspection arm of the government is the Electronic

Services Division of the Nevada Gaming Commission and State Gaming Control Board (http://gaming.nv.gov/esd_main.htm).

22. R&G Associates, LLC, "Parallel Monitoring Program, Summary Report,"19, April 2004, http://www.ss.ca.gov/elections/ks_dre_papers/Parallel _Monitoring_Summary_Report.pdf, See especially pp. 17–18.

23. R&G Associates, LLC, "Parallel Monitoring Program, Report of Findings", 30 November 2004, http://ss.ca.gov/elections/november2004_pmp _report.pdf. See especially p. 27.

24. Secretary of State, "Supplemental Report for Merced County," 10 December 2004, http://ss.ca.gov/elections/pmp_merced.pdf.

25. See the discussion of these "glitches" at http://electionupdates.caltech .edu/2006_11_05_archive.html.

26. For example, see the discussion of the inclusion of 700 previously disallowed provisional ballots in King County at http://en.wikipedia.org/wiki/ Washington_gubernatorial_election,_2004.

27. See the 2000 exit poll report by CNN at http://www.cnn.com/ELECTION/ 2000/results/index.epolls.html. There, African Americans supported Gore over Bush by a wide margin, 90 to 9 percent. Individuals with incomes less than $15,000 supported Gore over Bush, 57 to 37 percent; those with incomes between $15,000 and $30,000 also supported Gore over Bush by a substantial margin in the exit poll data, 54 to 41 percent.

28. The log transformation used in the first two regression models allows us to estimate the residual vote model and address the fact that the residual vote measure has an unusual distribution, with a lower bound at zero and with much of the mass of the distribution clustered just above zero. Such an unusual distribution can produce problems when ordinary-least squares regression models are estimated model parameters; the use of the log transformation of the residual vote measure helps to smooth the distribution to make the data more consistent with the basic assumptions of least-squares regression.

29. The results in the table show a very high degree of consistency across the four model specifications, which means that the basic results about the impacts of voting technologies on residual vote rates in these data are robust with respect to model specification. The major differences across specifications are that we see a very important improvement in model fit when we use the county-year fixed effects specification, relative to the state-year specification, regardless of whether we use the logged or nonlogged version of the residual vote dependent variable.

NOTES TO CHAPTER 7

1. See http://www2.standardandpoors.com/spf/pdf/index/500factsheet.pdf for a briefing of this stock index, and http://www2.standardandpoors.com/spf/pdf/ index/062304_US.pdf 23 June 2004, for the original announcement that Gilead was being added to the S&P 500.

2. According to Gilead's 2005 10-K filing with the Securities and Exchange Commission, total revenues in 2004 were $1.3 billion—rising to over $2 billion in

2005 (http://www.sec.gov/Archives/edgar/data/882095/000119312506045128/
d10k.htm#tx13025_10).

3. For example, see Marcia Angell, "The Truth about the Drug Companies,"
New York Review of Books, 15 July 2004, http://www.nybooks.com/articles/
17244.

4. This billion dollar estimate was developed by the Caltech/MIT Voting
Technology Project in 2001, and is discussed in its 2001 report. No doubt that
election administration expenses have risen, especially after the passage of the
Help America Vote Act in 2002 and the various efforts in many states to invest
resources in election administration. There has been little study of election
administration finance; see Hill, 2006, for a rare exception.

5. Richard L. Hasen, Testimony of Richard L. Hasen, William H. Hannon
Distinguished Professor of Law, Loyola Law School, before the Commission on
Federal Election Reform (Carter-Baker Commission), 18 April 2005 Hearing,
American University, Washington, D.C., http://www.american.edu/ia/cfer/0418
test/ hasen.pdf. Also see Hasen's (2005) study of election administration.

6. See the text of HR 550 at http://www.holt.house.gov/pdf/HR percent20550
percent20VCIA.pdf.

7. See R. Michael Alvarez, Thad E. Hall, and Morgan Llewellyn, "American
Confidence in Electronic Voting and Ballot Counting, a Pre-Election Update,"
3 November 2006, http://www.annenberg.edu/files/2006-Voter-Confidence-Survey.
pdf. For detailed analysis of our survey research on voter confidence, see
Alvarez, Llewellyn, and Hall forthcoming a.

8. In addition to our study with Llewellyn on voter confidence (cited in h.7),
we have also studied public opinion regarding election governance (Alvarez,
Llewellyn, and Hall forthcoming b) and voter registration reform (Alvarez,
Llewellyn, and Hall 2007.

9. The primary exception to this were opinion polls conducted in early 2001;
see R. Michael Alvarez, Ann Crigler, Marion Just, Edward McCaffery, and
Robert Sherman, "American Opinion on Election Reform," 10 May 2001,
http://survey.caltech.edu/reform4.pdf. For analysis of this same survey data, see
Crigler, Just and Buhr 2004.

10. International Communications Research (ICR) describes the EXCEL
omnibus survey methodology as follows: "Each EXCEL survey consists of a
minimum of 1,000 interviews, $\frac{1}{2}$ with men and $\frac{1}{2}$ with women. EXCEL uses a
fully-replicated, stratified, single-stage random-digit-dialing (RDD) sample of
telephone households. Sample telephone numbers are computer generated and
loaded into online sample files accessed directly by the CRT system. Within each
sample household, one adult respondent is randomly selected using a computerized procedure based on the 'Most Recent Birthday Method' of respondent selection." See http://www.icrsurvey.com/Omni_Excel_ main.html for more details
about the ICR EXCEL methodology.

11. See Alvarez et al., "American Opinion on Election Reform"; see also
Crigler, Just, and Buhr 2004.

12. The survey report available to the public describes the introductory question,
but not the various voting system descriptions, as presented to survey
respondents. Respondents were also asked to evaluate optical scanning voting

technologies in precincts, optical scanning for voting by mail, and Internet voting. See M. Glenn Newkirk, "US Public Opinion toward Voting Technologies," 1 March 2004, http://www.infosentry.com/US_Public_Opinion_Toward_Voting_Technology_20040301.pdf.

13. The press release describing the survey data is located at http://company.findlaw.com/pr/2004/090704.electronicvoting.html. The release does not provide the precise wording of the survey questions making direct comparisons to our survey questions difficult.

14. For more detailed discussion about how information might reduce public uncertainty, see Alvarez 1997.

15. http://www.democracysystems.com/docs/press_release_nevada_study.pdf.

16. http://www.lombardoconsultinggroup.com/docs/nvvotersurvey.pdf.

17. The other randomly selected half of the second-wave sample was asked a series of questions about election governance. Our initial report regarding opinions about election governance is "Public Attitudes about Election Governance," June 2005, http://votingtechnologyproject.org/media/documents/election%20governance%20report_FINAL.pdf.

18. See "Voter-Verified Paper Audit Trail Laws and Regulations (as of 9/20/06)," produced by Electionline.org, http://electionline.org/Default.aspx?tabid=290.

19. For example, in early voting in Los Angeles County, where voters were using the Diebold AccuVote TSx equipped with a voter-verified paper audit device, we observed four of twenty early voters ask about their paper audit trail. For example, voters asked questions like, "Where's my copy of the ballot?" "Where did it go?" and "I'm disappointed I didn't get a printed ballot." In each case the poll workers on hand provide the voters with an explanation of the purpose of the VVPAT explain and that they don't get a copy of their voted ballot (see http://electionupdates.caltech.edu/2006/10/additional-observation-of-early-voting.html).

20. The Election Data Services (EDS) was issued on 2 October 2006: "Almost 55 Million, or One-Third of the Nation's Voters, Will Face New Voting Equipment in 2006 Election" (http:/electionupdates.caltech.edu/NR_VoteEquip_Nov-2006wTables.pdf).

21. There is some difficulty in knowing what type of voting technology some voters used because localities with precinct electronic voting have paper-based optical scan voting.

22. A preliminary analysis of the exit poll data and information about the exit poll methodology are available in the report, "November 2006 General Election Paper Trail Exit Poll Study," http://electionupdates.caltech.edu/UGA_Study_VVPAT_Nov_2006.pdf.

23. The Georgia legislature originally authorized these pilot tests as way to study whether the entire state should adopt a VVPAT system, and in the near future hearings will be held in Georgia to evaluate the pilot project and to determine whether the legislature should proceed with legislation mandating the VVPAT in Georgia.

NOTES TO CHAPTER 8

1. See, for example, the risk model outlined in Flyvbjerg 2003.

2. United States Department of Justice, Office of Justice Programs, National Institute of Justice, "Chemical Facility Vulnerability Assessment Methodology," no date (study conducted by the Sandia National Laboratories, Albuquerque, NM).

3. Ibid., 16.

4. RABA Innovative Solution Cell, "Trusted Agent Report: Diebold AccuVote-TS Voting System," 20 January 2004, pp. 5, 11, http:www.raba.com/press/TA_Report-Accuvote.pdf.

5. This is exactly the argument that was advanced by David Dill before the Election Assistance Commission in its 28 July 2005 hearing in Pasadena, California (www.eac.gov.)

6. Interview with Kathy Rogers, Georgia Director of Elections, 29 September 2005.

7. See Dana DeBeauvoir, "Method for Developing Security Procedures in a DRE Environment," http://www.co.travis.tx.us/county_clerk/election/pdfs/NIST _ paper_051005.pdf.

8. The Travis County model was given the "2005 Best Professional Practices Award" by the Election Center.

9. R&G Associates, LLC *Parallel Monitoring, California General Elections: Report of Findings,* (Folsom, CA, 30 November 2004), pp. 2–3.

10. FDA Staff Manual Guides, vol. III, *Quality System Framework for Internal Activities,* effective 1 September 2004, http://www.fda.gov/smg/vol13/2000/2020.html.

11. In an interesting MIT Master of Science thesis, Tomer Posner 2005 studied a different quality management approach, and how it was applied in the development and deployment of the Brazilian voting system. As far as we are aware, this is the only existing effort to apply quality management principles to the field of election administration.

12. Hall, Monson, Patterson, 2006, www.utah.edu/IPIA.

13. Much has been written on "Six Sigma"; a good starting point is the Six Sigma Web site, a portal to all things Six Sigma (http://www.isixsigma.com/).

BIBLIOGRAPHY

Alden, Robert. 1952a. "30% of Soldiers in Korea Voting." *New York Times,* 3 November, 1.

———. 1952b. "Stevenson Leads by 2–1 in Poll of 500 U.S. Army Men in Korea." New York Times, 1 November, 1.

Alvarez, R. Michael. 1997. *Information and Elections.* Ann Arbor: University of Michigan Press.

———. 2005a. "Precinct Voting Denial of Service." Prepared for NIST "Threats to Voting Systems" Workshop. Caltech/MIT Voting Technology Project Working Paper 39.

———. 2005b. "Threats to Statewide Voter Registration Systems." Paper prepared for NIST "Threats to Voting Systems Workshop." Caltech/MIT Voting Technology Project Working Paper 40.

Alvarez, R. Michael, Stephen Ansolabehere, and Charles Stewart III. 2005. "Studying Elections: Data Quality and Pitfalls in Measuring of Effects of Voting Technologies." *Policy Studies Journal* 33 (1): 15–24.

Alvarez, R. Michael, Melanie Goodrich, Thad E. Hall, D. Roderick Kiewiet, and Sarah M. Sled. 2004. "The Complexity of the California Recall Election." *PS: Political Science and Politics* 37 (January): 23–27.

Alvarez, R. Michael, and Thad E. Hall. 2004. *Point, Click, and Vote.* Washington, D.C.: Brookings Institution Press.

———. 2005a "Rational and Pluralistic Approaches to HAVA Implementation: The Cases of Georgia and California." *Publius: The Journal of Federalism* 35: 559–77.

———. 2005b. *The Next Big Election Challenge: Developing Electronic Data Transaction Standards for Election Administration.* IBM Center for the Business of Government, Washington, D.C.

Alvarez, R. Michael, Thad E. Hall, and Brian F. Roberts. Forthcoming. "Military Voting and the Law: Procedural and Technological Solutions to the Ballot Transit Problem." *Fordham Law Review.*

Alvarez, R. Michael, Thad E. Hall, and D. E. Sinclair. 2005. "Whose Absentee Votes Are Counted: The Variety and Use of Absentee Ballots in California." California Institute of Technology. Unpublished manuscript.

Alvarez, R. Michael, Morgan Llewellyn, and Thad E. Hall. 2007. "How Hard Can It Be: Do Citizens Think It Is Difficult to Register to Vote?" *Stanford Law and Policy Review* 18: 382–409.

———. Forthcoming a. "Are Americans Confident Their Ballots Are Counted." *Journal of Politics.*

———. Forthcoming b. "Who Should Run Elections in the United States." *Policy Studies Journal.*

Alvarez, R. Michael, D. E. Sinclair, and Richard L. Hasen. 2006. "How Much Is Enough? The Ballot Order Effect and the Use of Social Science Research in Election Law Disputes." *Election Law Journal* 5(1): 40–56.

Alvarez, R. Michael, D. E. Betsy Sinclair, and Catherine Wilson. 2004. "Counting Ballots and the 2000 Election: What Went Wrong?" In *Rethinking the Vote: The Politics and Prospects of American Electoral Reform,* ed. Ann Crigler, Marian R. Just, and Edward J. Mc Caffery, 34–50. New York: Oxford University Press.

Anderson, Troy, and Beth Barrett. 2005. "State to Join Ballot Probe." *Daily News of Los Angeles,* 17 March, N1.

Ansolabehere, Stephen. 2002. "Voting Machines, Race and Equal Protection." *Election Law Journal* 1(1): 61-70.

Ansolabehere, Stephen, and Shanto Iyengar. 1995. *Going Negative: How Political Advertisements Shrink and Polarize the Electorate.* New York: Free Press.

Ansolabehere, Stephen, and Andrew Reeves. 2004. "Using Recounts to Measure the Accuracy of Vote Tabulations; Evidence from New Hampshire Elections, 1946–2002." Caltech/MIT Voting Technology Project, Working Paper 11.

Ansolabehere, Stephen, Jonathan Rodden, and James M. Snyder Jr. 2006. "Purple America." *Journal of Economic Perspectives* 20 (2): 97–118.

Ansolabehere, Stephen, and Charles Stewart III. 2005. "Residual Votes Attributable to Technology." *Journal of Politics* 67 (2): 365–89.

Associated Press. 2004. "7,000 SoCal Voters Reportedly Given Wrong Ballots." 9 March.

Bain, Henry M., Jr., and Donald S. Hecock. 1957. *Ballot Position and Voter's Choice.* Detroit: Wayne University Press.

Baird, Brian. 2004. "We Need to Read the Bills." *Washington Post,* 27 November, A31.

Barrett, Beth. 2005. "Voting Results Won't Be Disputed." *Daily News of Los Angeles,* 19 March, N1.

Barstow, David, and Don Van Natta Jr. 2001. "How Bush Took Florida: Mining the Overseas Absentee Vote." *New York Times,* 15 July, 1.

Baumgartner, Frank, and Bryan Jones. 1993. *Agendas and Instability in Public Policy.* Chicago: University of Chicago Press.

Beck, Ulrich. 1992. *Risk Society: Toward a New Modernity.* London: Sage.

———. 1998. "Politics of Risk Society." In *The Politics of Risk Society,* ed. Jane Franklin, 9–22. Cambridge: Polity Press.

———. 1999. *World Risk Society.* Malden, Mass: Blackwell.

Bensel, Richard Franklin. 2004. *The American Ballot Box in the Mid-Nineteenth Century.* New York: Cambridge University Press.

Bousquet, Steve. 2004. "GOP Flier Questions New Voting Equipment." *St. Petersburg Times,* 29 July.

Bower, Tom. 2002. "Ballot Blunder Admitted." *San Antonio Express-News,* 13 November, A-1.

Brady, Henry E. 2004. "Performance of Voting Systems on March 2, 2004." University of California, Berkeley.

Brady, Henry E., Justin Buchler, Matt Jarvis, and John McNulty. 2001. "Counting All the Votes: The Performance of Voting Technology in the United States." University of California at Berkeley. Unpublished Manuscript.

Brady, Henry E., Guy-Uriel Charles, Benjamin Highton, Martha Kropf, Walter R. Mebane Jr., and Michael Traugott. 2004. "Interim Report on Alleged Irregularities in the United States Presidential Election of 2 November 2004." National Research Commission on Elections and Voting, Social Science Research Consortium. http://www.vote.caltech.edu/media/documents/Interim Report122204-1.pdf.

Brown, JoAnne. 1990. "The Social Construction of Invisible Danger: Two Historical Examples." In *Nothing to Fear: Risks and Hazards in American Society,* ed. Andrew Kirby, 39–52. Tucson: University of Arizona Press.

Buchler, Justin, Matthew Jarvis, and John E. McNulty. 2004. "Punch Card Technology and the Racial Gap in Residual Votes." *Perspectives on Politics.* 2 (3): 517–24.

Bullock, Charles S., III, and M. V. Hood III. 2002. "One Person-No Vote; One Vote; Two Votes: Voting Methods, Ballot Types, and Undervote Frequency in the 2000 Presidential Election." *Social Science Quarterly.* 83: 981–93.

Bynum, Russ. 2000. "Florida Courts Have Picked Election Winner Before." Associated Press, 28 November.

Caltech/MIT Voting Technology Project. 2001. "Voting: What Is, What Could Be. Manuscript, Pasadena, Calif.

———. 2004a. "Voting Machines and the Underestimate of the Bush Vote." Monograph, Pasadena, Calif.

———. 2004b. "On the Discrepancy between Party Registration and Presidential Vote in Florida." Monograph, Pasadena, Calif.

Celeste, Richard F., Dick Thornburgh, and Herbert Lin. 2006. Asking the Right Questions about Electronic Voting. Washington, D.C.: National Academy Press.

Cone, Marla. 2005. "Europe's Rules Forcing US Firms to Clean Up." *Los Angeles Times*, 16 May.

Cox, Cathy. 2001. *The 2000 Election: A Wake-Up Call. Report to the Governor and Members of the General Assembly.* Office of the Secretary of State, Atlanta, Georgia.

Crigler, Ann N., Marion R. Just, and Tami Buhr. 2004. "Cleavage and Consensus: The Public and Election Reform." In *Rethinking the Vote: The Politics and Prospects of American Electoral Reform,* ed. Ann N. Crigler, Marion R. Just, and Edward J. McCaffery, 151–66. New York: Oxford University Press.

Edison/Mitofsky. 2005. "Evaluation of Edison/Mitofsky Election System 2004." Prepared by Edison Media Research and Mitofsky International for the National Election Pool (NEP). http://www.exit-poll.net/election-night/ EvaluationJan192005.pdf.

Electoral Commission. 2003. "The Shape of Things to Come." London.

FDA. *Pharmaceutical [Current Good Manufacturing Practices] for the 21st Century.* http://www.fda.gov/cder/gmp/.

Feldman, A. J., J. A. Halderman, and E. W. Felten. 2006. "Security Analysis of the Diebold AccuVote-TS Voting Machine." Princeton University Working Paper.

Fiorina, Morris P. with Samuel J. Abrams and Jeremy C. Pope. 2005. *Culture War? The Myth of A Polarized America*. New York: Pearson/Longman.

Flyvbjerg, Bent. 2003. *Megaprojects and Risks*. New York: Cambridge University Press.

Fortier, John C., and Norman J. Ornstein. 2004. "If Terrorists Attacked Our Presidential Elections." *Election Law Journal* 3(4): 597–612.

Franklin, J. 1998. *The Politics of Risk Society*. Cambridge: Polity Press.

Freeman, Steve, and Joel Bleifuss. 2006. *Was the 2004 Presidential Election Stolen? Exit Polls, Election Fraud, and the Official Count*. New York: Seven Stories Press.

Frisina, Laurin, Michael C. Herron, James Honaker, and Jeffrey B. Lewis. 2007. "Ballot Formats, Touchscreens, and Undervotes: A Study of the 2006 Midterm Elections in Florida." Dartmouth College. Unpublished manuscript.

General Accounting Office (GAO). 1998. *Executive Guide: Information Security Management*. GAO/AIMD-98-68. May. Washington, D.C.

———. 2001a. "Elections: Issues Affecting Military and Overseas Absentee Voters." GAO-01-704T. 9 May. Washington, D.C.

———. 2001b. "Statistical Analysis of Factors that Affected Uncounted Votes in the 2000 Presidential Election." GAO-02-122. October. Washington, D.C.

———. 2004. "Operation Iraqi Freedom: Long Standing Problems Hampering Mail Delivery Need to Be Resolved." GAO-04-484. April. Washington, D.C.

Getches, Catherine. 2004. Letter to the Editor. *Los Angeles Times,* 10 March.

Giddens, Anthony. 1998. "Risk Society, The Context of British Politics." In *The Politics of Risk Society,* 23–34. ed. J. Franklin, Cambridge: Polity Press.

Goodnough, Abby. 2004. "Lost Record of Vote in '02 Florida Race Raises '04 Concern." *New York Times*, 27 July, A1.

Goodrich, Melanie. 2004. "19th Century Ballot Reform in California: A Study of the Huntington Library's Political Ephemera Collection." Caltech/MIT Voting Technology Project Working Paper 1.

Grigg, Delia. 2005. "Measuring the Effect of Voting Technology on Residual Votes." California Institute of Technology. 16 May.

Gumbel, Andrew. 2005. *Steal This Vote: Elections and the Rotten History of Democracy in America*. New York: Nation Books.

Hall, Thad E. 2003. "Public Participation in Election Management: The Case of Language Minority Voters." *American Review of Public Administration* 33 (4): 407–22.

———. 2002. "LA Story: The 2001 Election." Century Foundation. 1 January.

Hall, Thad E., Quin Monson, and Kelly Patterson. 2006. "The Human Dimension of Elections: How Poll Workers Shape Public Confidence in Elections." IPIA Working Paper.

Halvorsen, Robert, and Raymond Palmquist. 1980. "The Interpretation of Dummy Variables in Semilogarithmic Equations." *American Economic Review* 70 (3): 474–75.

Hamner, Michael J., and Michael W. Traugott. 2004. "The Impact of Voting by Mail on Voter Behavior." *American Politics Research,* 32: 375–405.

Harrington, Mark. 2003. "Internet Election Company Sells Assets to Business Partner." *Newsday*, 2 July.

Harris, Bev. 2004. "Black Box Voting: Ballot Tampering in the 21st Century." Renton, WA: Talion Publishing.

Harris, Joseph P. 1934. *Election Administration in the United States*. Washington, D.C.: Brookings Institution.

Harris, P., and N. O'Shaughnesy. 1997. "BSE and Marketing Communication Myopia." *Risk, Decision, and Policy* 2(1): 29–39.

Hasen, Richard. 2002. "Bush v. Gore and the Future of Equal Protection Law in Elections." *Florida State University Law Review* 29: 377–406.

———. 2005. "Beyond the Margin of Litigation: Reforming U.S. Election Administration to Avoid Electoral Meltdown." *Washington and Lee Law Review* 63: 937–99.

Herndon, Ray, and Stuart Pfeifer. 2004. "The State: 7,000 Orange County Voters Were Given Bad Ballots." *Los Angeles Times*, 9 March, A-1.

Herron, Michael, and Jeffrey B. Lewis. 2006. "From Punchcards to Touchscreens: Some Evidence from Pasco County, Florida on the Effects of Changing Voting Technology." Dartmouth College Working Paper.

Herron, Michael, and Jasjeet Sekhon. 2001. "Overvoting and Representation: An Examination of Overvoted Presidential Ballots in Broward and Miami-Dade Counties." Harvard University. Unpublished manuscript.

Hill, Sarah A. 2006. "Election Finance in California." California Institute of Technology. Unpublished manuscript.

Ho, David E., and Kosuke Imai. 2005. "The Impact of Partisan Electoral Regulation: Ballot Effects from the California Alphabet Lottery, 1978–2002." Princeton University. Unpublished manuscript.

Hoffman, Kathy Barks. 2003. "Internet Voting Moves Step Closer in Michigan Democratic Caucuses." *Deseret News,* 1 November.

———. 2004a. "Voters Find Some Caucus Sites Moved; Other Sites Run Out of Ballots." Associated Press, 7 February.

———. 2004b. "Detroit Caucus Sites to Stay Open Extra Two Hours." Associated Press, 7 February .

Imai, Kosuke, and Gary King. 2004. "Did Illegally Counted Absentee Ballots Decide the 2000 U.S. Presidential Election?" *Perspectives on Politics* 2 (3): 537–50.

Inglehart, Ronald. 1984. "The Fear of Living Dangerously: Public Attitudes toward Nuclear War." *Public Opinion* 7: 41–44.

Iyengar, Shanto, and Donald R. Kinder. 1989. *News That Matters: Television and American Opinion*. Chicago: University of Chicago Press.

Jefferson, David, Aviel D. Rubin, Barbara Simons, and David Wagner. 2004. "A Security Analysis of the Secure Electronic Registration and Voting Experiment (SERVE)." 21 January. httpa://www.servesecurityreport.org/paper. pdf.

Jelinek, Pauline. 2006. "Web Site Improved to Help Overseas Voters: Many Troops Said They Didn't Vote Because They Didn't Know How." *Houston Chronicle*, 8 September, 24.

Karatnycky, Adrian. 2000. "Ukraine's Orange Revolution." *Foreign Affairs*. 84 (2): 35–52.

Kasperson, Roger E., and Jeanne X. Kasperson. 1996. "The Social Amplification and Attenuation of Risk." *Annals of the American Academy of Political and Social Science 545*: 95–105.

Kennedy, Peter E. 1981. "Estimation with Correctly Interpreted Dummy Variables in Semilogarithmic Equations." *American Economic Review* 71 (4): 801.

Keyssar, Alexander. 2000. *The Right to Vote: The Contested History of Democracy in the United States.* New York: Basic Books.

Knack, Stephen, and Martha Kropf. 2003. "Invalidated Ballots in the 1996 Presidential Election: A County-Level Analysis." *Journal of Politics* 65 (3): 881–97.

Kohno, Tadayoski, Adam Stubblefield, Aviel Rubin, and Dan Wallach. 2004. "Analysis of an Electronic Voting System." IEEE Symposium on Security and Privacy 2004. Piscataway. N.J.: IEEE Society Press.

Koppell, Jonathan G. S., and Jennifer A. Steen. 2004. "The Effects of Ballot Position on Election Outcomes." *Journal of Politics* 66: 267–81.

Krugman, Paul. 2004. "Saving the Vote." *New York Times,* 17 August.

Lanoutte, William. 1990. "How Atomic Agency Managed the News in the Early Years." *Newsletter of the National Association of Science Writers* 38: 1–3.

Liberto, Jennifer. 2004. "Two Blind Voters Say Privacy Was Violated." *St. Petersburg Times,* 9 November, 6.

Ludington, Arthur. 1911. "Progress of Short Ballot Movement." *American Political Science Review* 5 (1): 79–83.

Magleby, David B. 1984. *Direct Legislation; Voting on Ballot Propositions in the United States.* Baltimore: Johns Hopkins University Press, 1984.

Martin, Boyd A. 1945. "The Service Vote in the Elections of 1944." *American Political Science Review* 39 (4): 720–32.

McCullough, B. D., and Florenz Plassman. 2004. "A Partial Critique of Hout, Mangels, Carlson and Best's 'The Effect of Electronic Voting Machines on Change in Support for Bush in the 2004 Florida Elections.'" National Research Commission on Elections and Voting, A Social Science Research Council Project. http://election04.ssrc.org/research/critique-of-hmcb.pdf.

McGreevy, Patrick. 2005. "City Clerk Defends Over-Marking of Mayoral Ballots." *Los Angeles Times* 22 March, B-3.

Mercuri, Rebecca. 2000. "Electronic Vote Tabulation: Checks and Balances." Ph.D. dissertation, University of Pennsylvania.

Merriam, Charles E., and Louise Overacker. 1928. *Primary Elections.* Chicago: University of Chicago Press.

Merzer, Martin, Joni James, and Alfonso Chardy. 2002. "In Florida, Confusion Reigns in Sequel to 2000 Election." *Miami Herald,* 11 September.

Miami-Dade County Office of Inspector General. 2002. "OIG Inquiry Into Circumstances Surrounding the September 10, 2002, Election in Miami-Dade County." Report, 20 September.

———. 2003. Cover Memorandum and Summary of the OIG's Final Report of the Miami-Dade County Voting Systems. Contract RFP No. 326, 20 May.

Miller, Joanne E., and Jon A. Krosnick. 1998. "The Impact of Candidate Name Order on Election Outcomes." *Public Opinion Quarterly* 62: 291–330.

Mock, Jennifer. 2004. "Rep. Rodriquez Alleges Fraud after Recount Reverses Outcome and Hands Primary Win to His Rival." *CQ Weekly,* 10 April, 858.

Moses, Alexandra R. 2004. "Party Says Just Over 46,000 People Voted Online in State Democratic Caucuses." Associated Press, 7 February.

Mythen, Gabe. 2004. *Ulrich Beck: A Critical Introduction to The Risk Society.* Sterling, V.: Pluto Press.

Neumann, Peter. 1990. "Risks in Computerized Elections." Inside Risks 5. *Communications of the ACM* 33, 11: 33–170.

New York Times. 1923."The Only Test." 3 November, 12.

Orlov, Rick. 2005. "Slow Ballot Count Defended." *Daily News of Los Angeles,* 24 March, N3.

Patterson, Thomas E. 2003. *The Vanishing Voter: Public Involvement in an Age of Uncertainty.* New York: Vintage Books.

Pickler, Nedra. 2003a. "Internet Voting Stirs Debate in Michigan." 20 November. http://www.usatoday.com/tech/news/techpolicy/2003-11-20-net-voting-mich_ x.htm.

———. 2003b. "Democrats OK Michigan Internet Voting." 23 November. http: // www.mercurynews.com/mld/mercurynews/news/politics/7328526.htm?1c.

Pidgeon, Nick, Roger Kasperson, and Paul Slovic. 2003. The Social Amplification of Risk. Cambridge: Cambridge University Press, 2003.

Posner, Tomer. 2005. "Application of Lean Management Principles to Election Systems.", M.S. thesis, Massachusetts Institute of Technology.

Riker, William. 1986. *The Art of Political Manipulation.* New Haven: Yale University Press.

Roig-Franzia, Manuel. 2004. "Vote Data Lost in Space—Next Door; Elections Chief Tries to Explain Latest Twist in Fla. Series." *Washington Post,* 31 July, A7.

Roth, S. K. 1998. "Disenfranchised by Design: Voting Systems and the Election Process." *Information Design Journal* 9 (1): 1–8.

Rothman, Stanley, and S. Robert Lichter. 1982. "The Nuclear Energy Debate: Scientists, the Media, and the Public." *Public Opinion* 5: 47–52.

Saletan, William. 2003. *Bearing Right: How Conservatives Won the Abortion War.* Berkeley: University of California Press.

Saltman, Roy. 1988. "Accuracy, Integrity, and Security in Computerized Vote-Tallying," NBS Special Publication: 500–158. Washington, D.C.: Institute for Computer Sciences and Technology, National Bureau of Standards.

———. 2006. *The History and Politics of Voting Technology: In Quest of Integrity and Public Confidence.* New York: Palgrave-Macmillan.

Seelye, Katharine Q. 2004. "Michigan's Online Ballot Spurs New Strategies for Democrats." *New York Times,* 10 January.

Sekhon, Jasjeet. Forthcoming. "Data Troubles: Explaining Discrepancies between Official Votes and Exit Polls in the 2004 Presidential Election." *Chance.*

Selker, Ted. Matt Hockenberry, John Goler, and Shawn Sullivan. 2005. "Orienting Graphical User Interfaces Reduce Errors: The Low Error Voting Interface." Caltech/MIT Voting Technology Project Working Paper 23.

Seybold, Patricia B. 2001. "Get Inside the Lives of Your Customers." *Harvard Business Review* 79: 80–89.

Shamos, Michael Ian. 1993. "Electronic Voting—Evaluating the Threat." Proceedings of the Third ACM Conference on Computers, Freedom and Privacy, San Francisco, March.

———. 2004. "Paper v. Electronic Voting Records—An Assessment." Proceedings of the 14th ACM Conference on Computers, Freedom and Privacy, Berkeley.

———. 2005. "Paper v. Electronic Voting Records—An Assessment." 27 November. http://euro.ecom.cmu.edu/people/faculty/mshamos/paper.htm.

Sinclair, D. E., and R. Michael Alvarez. 2004. "Who Overvotes. Who Undervotes, Using Punchcards? Evidence from Los Angeles County." *Political Research Quarterly* 57(March): 15–25.

Sled, Sarah M. 2003. "Vertical Proximity Effects in the California Recall." Caltech/MIT Voting Technology Project Working Paper 8.

Slovic, Paul. 2000. *The Perception of Risk*. London: Earthscan.

Stephanie, Miranda. 1968. "Absentee Ballots." *New York Times*, 10 November.

Stewart, Charles, III. 2004. "The Reliability of Electronic Voting Machines in Georgia." Caltech/MIT Voting Technology Project Working Paper 20.

———. 2005. "Residual Vote in the 2004 Election." Caltech/MIT Voting Technology Project Working Paper, version 2.3.

———. 2006. "Changes in the Residual Vote Rates between 2000 and 2004." *Election Law Journal* 5(2): 158–69.

Stone, Deborah. 1988. *Policy Paradox and Political Reason*. Glenview, Ill.: Scott, Foresman.

———. 1989. "Causal Stories and the Formation of Policy Agendas." *Political Science Quarterly* 104: 281–300.

Sunstein, Cass. 2005. *Laws of Fear: Beyond the Precautionary Principle*. New York: Cambridge University Press.

Tampa Tribune. 2001. "Touch-Screens Get Vote of Confidence." 15 May, 2.

———. 2004. "Ongoing Voting Glitches Demand Extra Attention." 29 October, 18.

Terkildsen, Nayda, and Frauke Schnell. 1997. "How Media Frames Move Public Opinion: An Analysis of the Women's Movement." *Political Research Quarterly*. 50 (4): 879–900.

Therolk, Garrett. 2004. "Computer Lost Vote Results in Miami." *Tampa Tribune*, 29 July, Metro-1.

Tokaji, Daniel P. 2004. "The Paperless Chase: Electronic Voting and Democratic Values." *Fordham Law Review*. 73(4): 1711–1836.

Tolbert, Caroline J., John A. Grummel, and Daniel A. Smith. 2001. "The Effects of Ballot Initiatives on Voter Turnout in the American States." *American Politics Research* 29 (6): 625–48.

Tomz, Michael, and Robert P. van Houweling. 2003. "How Does Voting Equipment Affect the Racial Gap in Voided Ballots?" *American Journal of Political Science* 47 (1): 46–60.

Traugott, Michael, Benjamin Highton, and Henry E. Brady. 2005. "A Review of Recent Controversies Concerning the 2004 Presidential Election Exit Polls." New York: National Research Commission on Elections and Voting, Social Science Research Council.

van Garderen, Jan Kees, and Chandra Shat. 2002. "Exact Interpretation of Dummy Variables in Semilogarithmic Equations." *Econometrics Journal* 5: 149–59.

Wand, Jonathan. 2004. "Evaluating the Impact of Voting Technology on the Tabulation of Voter Preferences: The 2004 Presidential Election in Florida." Working paper, Stanford University.

Weart, Spenser. 1988. *Nuclear Fear: A History of Images.* Cambridge, Mass: Harvard University Press.

Weeks, George. 2004. "Michigan Caucus Provides Internet Voting Test." *Detroit News,* 6 February.

Wolter, K., D. Jergovic, W. Moore, J. Murphy, and C.O'Muircheartaigh. 2003. "Reliability of the Uncertified Ballots in the 2000 Presidential Election in Florida." *American Statistician* 57 (1): 1–14.

INDEX